DAVID G. CAMPBELL

The Crystal Desert

Summers in Antarctica

A MARINER BOOK

HOUGHTON MIFFLIN COMPANY

BOSTON · NEW YORK

For Jean, Karen, and Tatiana

First Mariner Books edition 2002
Copyright © 1992 by David G. Campbell

Visit our Web site: www.houghtonmifflinbooks.com.

Library of Congress Cataloging-in-Publication Data
Campbell, David G.
  The crystal desert : summers in Antarctica / David G. Campbell.
    p.   cm.
  Includes bibliographical references and index.
  ISBN 0-395-58969-X   ISBN 0-618-21921-8 (pbk.)
    1. Natural history — Antarctic regions. 2. Summer — Antarctic
regions. 3. Campbell, David G. — Journeys — Antarctic regions.
4. Antarctic regions — Description and travel. I. Title.
QH84.2.C36 1992 92-10583
508.98'9 — dc20 CIP

Printed in the United States of America

Book design by Robert Overholtzer
Maps by Jacques Chazaud

QUM 10 9 8 7 6 5 4 3 2

Houghton Mifflin Literary Fellowship Winners

E. P. O'Donnell, *Green Margins*
Dorothy Baker, *Young Man with a Horn*
Robert Penn Warren, *Night Rider*
Joseph Wechsberg, *Looking for a Bluebird*
Ann Petry, *The Street*
Elizabeth Bishop, *North & South*
Anthony West, *The Vintage*
Arthur Mizener, *The Far Side of Paradise*
Madison A. Cooper, Jr., *Sironia, Texas*
Charles Bracelen Flood, *Love Is a Bridge*
Milton Lott, *The Last Hunt*
Eugene Burdick, *The Ninth Wave*
Philip Roth, *Goodbye, Columbus*
William Brammer, *The Gay Place*
Clancy Sigal, *Going Away*
Edward Hoagland, *The Cat Man*
Ellen Douglas, *A Family's Affairs*
John Stewart Carter, *Full Fathom Five*
Margaret Walker, *Jubilee*
Berry Morgan, *Pursuit*
Robert Stone, *A Hall of Mirrors*
Willie Morris, *North Toward Home*
Georgia McKinley, *Follow the Running Grass*
Elizabeth Cullinan, *House of Gold*
Edward Hannibal, *Chocolate Days, Popsicle Weeks*
Helen Iglesias, *How She Died*
Henry Bromell, *The Slightest Distance*
Julia Markus, *Uncle*
Jean Strouse, *Alice James*
Patricia Hampl, *A Romantic Education*
W. P. Kinsella, *Shoeless Joe*
David Payne, *Confessions of a Taoist on Wall Street*
Ethan Canin, *Emperor of the Air*
David Campbell, *The Crystal Desert*
Ashley Warlick, *The Distance from the Heart of Things*

Contents

Acknowledgments

No human can survive alone in the Antarctic, and one's companions in that hostile continent become lifelong friends. I am deeply indebted to Dr. Renato George Ferreira Garcia for inviting me to conduct research at Comandante Ferraz, the Brazilian Antarctic station, for showing me the ropes, and for a fruitful scientific collaboration, which continues today. I would also like to thank Captain Antonio José Gomez Queiroz, Dr. Alexandre de Azavedo Dutra, Dr. Mirian Maria Ferreira Garcia, Dr. Edson Rodrigues, Dr. Claude de Broyer, Gautier Chapelle, Dr. Johann-Wolfgang Wägele, Dr. Hans Gerd Mers, Dr. Phan Van Ngan, Helena Guiro P. P. Coelho, and Ézero Izidorio Tardin.

I am grateful to Pat Wilcoxon, University of Chicago Libraries, for access to that wonderful collection. I thank G. Douglas, Librarian of the Linnaean Society of London; and William Mills, Librarian, and Robert Headland, Archivist, of the Scott Polar Research Institute, Cambridge, for providing records of the early exploration of the Antarctic Peninsula.

For making many helpful comments as to content and style of the manuscript, I thank Dr. Diane Ackerman, Charles Bassett, Lizzie Grossman, Dr. Charles Swithinbank, Dr. Jere Lipps, Dr. Susan Trivelpiece and, especially, Harry Foster and Peg Anderson, editors at Houghton Mifflin.

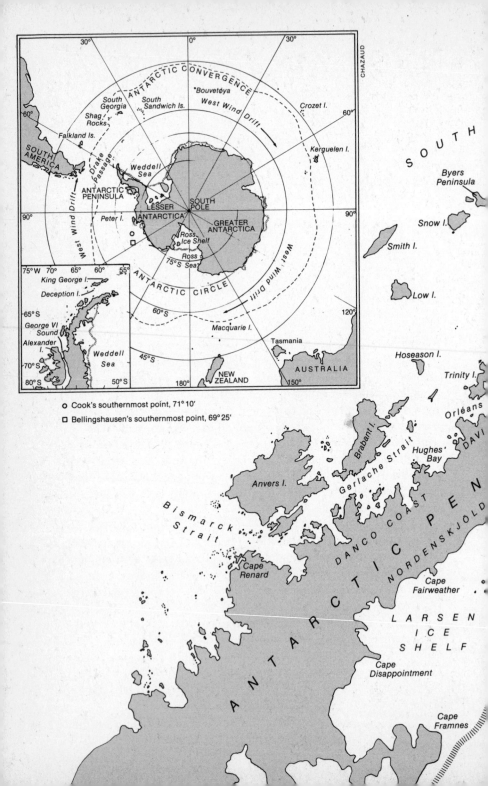

CHAZAUD

30°   0°   30°

ANTARCTIC CONVERGENCE

•Bouvetøya

West Wind Drift

Crozet I.

60°

South
Georgia

South
Sandwich Is.

60°

Shag
Rocks

Kerguelen I.

Falkland Is.

S O U T H

SOUTH
AMERICA

Drake Passage

Weddell
Sea

ANTARCTIC
PENINSULA

LESSER
ANTARCTICA

SOUTH
POLE

Byers
Peninsula

90°

West Wind Drift

Peter I.

GREATER
ANTARCTICA

90°

Snow I.

Ross
Ice Shelf

Smith I.

Ross
Sea

75°S

West Wind Drift

Low I.

ANTARCTIC CIRCLE

60°S

120°

75°W 70° 65° 60° 55°

King George I.

Hoseason I.

60°S

Deception I.

Macquarie I.

Trinity I.

65°S

Tasmania

Orléans

George VI
Sound

Weddell
Sea

45°S

AUSTRALIA

DAVI

Alexander
I.

Brabant I.

Gerlache Strait

Hughes
Bay

70°S

180°

NEW
ZEALAND

150°

80°S 50°S

Anvers I.

DANCO COAST

○ Cook's southernmost point, 71° 10'

□ Bellingshausen's southernmost point, 69° 25'

A N T A R C T I C   P E N

Bismarck
Strait

NORDENSKJÖLD

Cape
Renard

Cape
Fairweather

A   N   T   A   R   C

L A R S E N

I C E

S H E L F

Cape
Disappointment

Cape
Framnes

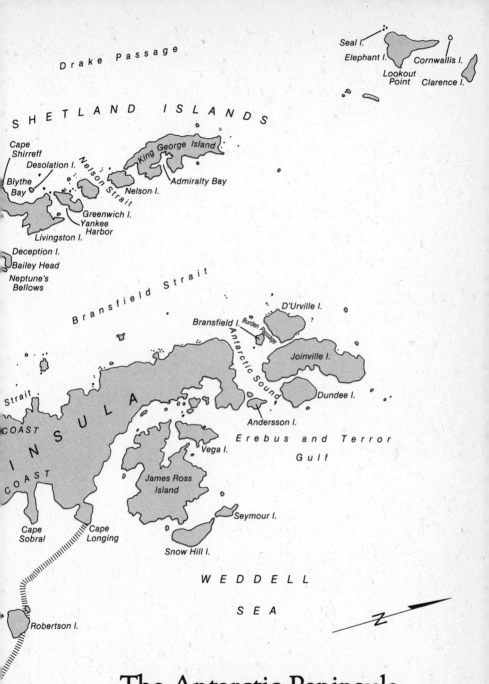

Drake Passage

SHETLAND ISLANDS

Cape Shirreff
Desolation I.
Nelson Strait
Blythe Bay
Nelson I.
King George Island
Admiralty Bay
Greenwich I.
Yankee Harbor
Livingston I.
Deception I.
Bailey Head
Neptune's Bellows

Seal I.
Elephant I.
Lookout Point
Cornwallis I.
Clarence I.

Bransfield Strait

Bransfield I.
Burden Passage
D'Urville I.
Joinville I.
Antarctic Sound
Dundee I.
Andersson I.

Strait
COAST
INSULA
COAST

Erebus and Terror Gulf

Vega I.
James Ross Island
Seymour I.
Cape Sobral
Cape Longing
Snow Hill I.

WEDDELL

SEA

Robertson I.

The Antarctic Peninsula and South Shetland Islands

0   100   200 km

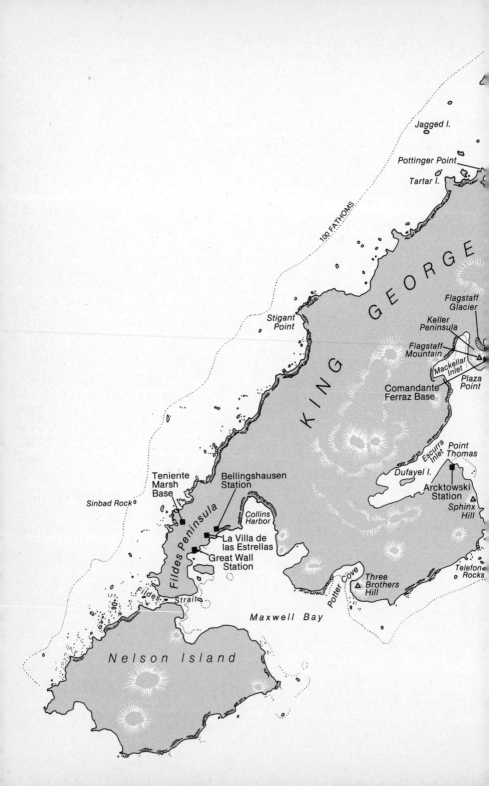

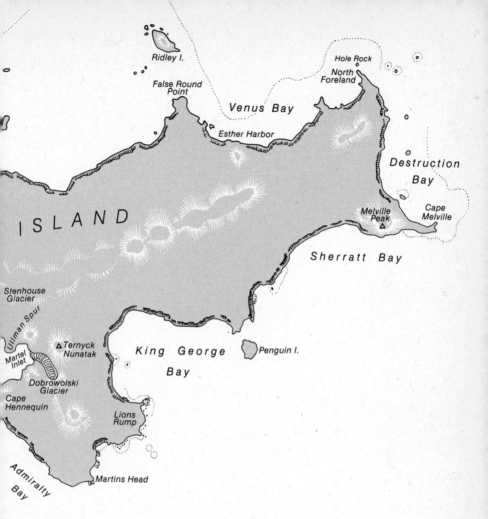

Ridley I.

False Round
Point

Hole Rock

North
Foreland

Venus Bay

Esther Harbor

Destruction
Bay

Melville
Peak △

Cape
Melville

I S L A N D

Sherratt Bay

Stenhouse
Glacier

Ullman Spur

△Ternyck
Nunatak

Martel
Inlet

King George

Bay

Penguin I.

Dobrowolski
Glacier

Cape
Hennequin

Lions
Rump

Admiralty
Bay

Martins Head

—Demay Point

King George Island

0   5   10   20 km

CHAZAUD

tremes of temperature are modulated by the sea. Explorers who have been to the frigid interior of the continent call the peninsula and its nearby islands the "Banana Belt" of Antarctica.[1] It rains frequently during the summer, and once, in late January, I watched the thermometer climb to 9° centigrade. The rest of the continent, ice-fast and arid, is a true desert and is mostly lifeless.

Northwest of the peninsula are the South Shetlands, an ocean-sculpted arc of islands, some with active volcanoes, that reminded the first homesick and frightened Scottish sealers of those treeless, windblown islands of the North Sea. The sealers named the new islands Clarence, Elephant, King George, Nelson, Greenwich, Livingston, and Snow, after various monarchs, captains, mammals, meteorological events, and hometowns. Set off from the archipelago is aptly named Deception Island, an active volcano with a secret caldera where ash-blackened snow mimics rock. These islands of ice and black basalt, now and then tinged russet or blue by oozings of iron or copper, rise over 600 meters. Their hearts are locked under deep glaciers, a crystal desert forever frozen in terms of our short life spans, but transient in their own time scale. Sometimes one sees only the cloud-marbled glacial fields, high in the sun above hidden mountain slopes and sea fog, Elysian plains that seem as insubstantial as vapor. The interiors of the glaciers, glimpsed through crevasses, are neon blue. Sliding imperceptibly on their bellies, the glaciers carve their own valleys through the rock, and when they pass over rough terrain they have the appearance of frozen rapids, which is in fact what they are, cascading at the rate of a centimeter a day. Sudden cold gusts, known as katabatic winds, tumble down their icefalls to the shore; sometimes the coast snaps from tranquility to tempest in just a few minutes. Just as quickly the glacial winds abate, and there is calm. Where they reach the sea, the glaciers give birth to litters of icebergs, which usually travel a short distance and, at the next low tide, run aground on hidden banks. Most of the ice-free land is close to shore, snuggled near the edge of the warm sea in places that are buffeted by both sea wind and land wind, where rain changes to snow and back. There is no plant taller than a lichen here, no animal larger than a midge — biological haiku. But on protected slopes, where the snow melts on warm summer days and glacial meltwater nourishes the soil, lichens and mosses dust the hills a pale gray-green, and the islands take on a tenuous verdancy.

The Pacific and Atlantic oceans meet at the South Shetland Islands. Indeed, all of the world's oceans mix in the Southern Ocean, the circumpolar sea that so absolutely isolates Antarctica from the other continents. Only a few small islands fleck this globe-girdling sea, and the westerly storms that orbit the Earth at these latitudes, unimpeded by land masses and always sucking energy from the sea, develop the anger of hurricanes. These zones are the "roaring forties" and "screaming fifties," which have commanded the respect and fear of sailors since the time of Francis Drake. Today satellite photographs show these low-pressure zones, spiraling clockwise, regularly spaced, separated by several hundred kilometers of calm sea — a flowered anklet on the planet Earth. But the Southern Ocean is a manic sea, and between the tempests there is tranquility and light. To the land-bound on the South Shetland Islands, the distance between cyclones is measured in time, three to five days apart.

If the bright ice and dark rock are the canvas of these desert islands, then light is the medium, and the Southern Ocean, ever fickle, often angry, is the artist. She swathes the islands in mist, or snow, or clarid sea-light, depending on her moods. Sometimes the sea rages for days, and you can lean against the wind, rubbery and firm. The wind lifts the round pebbles from the beach and flings them like weapons at hapless beachcombers. The cyclonic winds march around the compass, so at one moment they will herd the icebergs against the shore in a groaning cluster, and a few hours later waft them out to sea like feathers on a pond. At other times there will be a clammy calm, a disquieting purple grayness that smothers light and sound, punctuated only by the distant, muffled crack of a calving glacier. The days one anticipates are the tranquil ones that break the long captivity of cabin fever, when the sky is a transparent blue, and the deep, clear sea scintillates with shafts of sunlight.

King George Island lies in the middle of the South Shetland archipelago. Like the Antarctic Peninsula, 96 kilometers to the east across Bransfield Strait, it seems to bend slightly to the westerly currents. Ninety-five percent of the island is permanently covered with ice. The smooth domes of glaciers are the highest places on the island, all above 580 meters. Echo-sounding has proved that some of these glaciers have their feet at sea level. Along the shore the temperature is more or less constant, no more than 5° or 6° C above or below freezing, winter and summer. The seasonal and daily cycles of freeze and thaw create a disheveled landscape of wet landslides and cleaved rocks; some rocks are shattered in leaves, like sliced bread. There is precipitation three hundred days a year, and an equal probability of rain and snow during all months of the year. The northern shore of King George Island takes the full brunt of the wind and the sea; the breakers arrive unimpeded all the way from Tasmania. The coast is flecked by numerous rocky islets and scalloped by crescent bays. Many features are uncharted and unnamed. Others bear names of fancy: Sinbad Rock, Jagged Island, Tartar Island, Hole Rock, Venus Bay; or of tragedy: Destruction Bay.

On the north face of King George Island, the outer edge of Antarctica, cliffs of ice act as giant airfoils, pushing the sea wind up their faces onto the frozen, white plateau of the island. Once, during a blizzard, I walked along this shore and watched fat snowflakes fly skyward in seeming defiance of gravity. In the dark ocean below, the ricocheting swells converged on unseen banks, and the sea seemed to be spontaneously erupting. On clear days the summer sunlight refracts in these sea-lenses and the ocean appears to be glowing from within. In the half-moon bays below the cliffs, the volcanic basalt is ground to smooth pebbles. During the austral summer the seafaring flotillas of juvenile penguins come ashore to rest, and their white bellies look like stranded icebergs on the black beaches. The elephant seals, like huge grubs, wallow in their own oily excrement. If you walk amid the shuffle of broken algae and limpet shells, where the pebbles boom with each breaking swell and the icebergs ping and crackle offshore, you smell none of the familiar seashore odor of decomposition of more temperate climes; there are no beach flies or scuttling crabs. Everything remains frozen, immutable.

The southern, leeward shore of King George Island is not as rugged and steep as the north, and most ships arriving across the stormy Drake Passage from the tip of South America, 970 kilometers to the north, hie to shelter there. They skirt the North Foreland, on the eastern edge of the island closest to South America, and sail past Cape Melville, Sherratt Bay, Three Sisters Point, Penguin Island, and King George Bay, where in 1819 William Smith, commanding the brigantine *Williams,* planted the Union Jack and named the island after his distant and unknowing sovereign. And then a vast anvil-shaped bay opens to the northwest and invades the heart of the island. This is Admiralty Bay, a three-fingered fjord that is one of the safest anchorages in all of Antarctica. It is also perhaps the prettiest place in Antarctica. The bay splits into three deep inlets: Mackellar, which suffers a southern exposure and is sometimes chopped by storm and wave; Escurra, which is scoured by sudden katabatic winds; and Martel, which offers safe anchorage in winds from all directions. All three inlets terminate in glaciers that flow down from the heart of the island, nudging spongy moraines along their flanks, and during the summer cluttering the bay with icebergs. The bay itself was born of fire and ice. A tectonic fault line, where two plates scrape against each other, is buried five hundred meters deep. The flanks of the bay were carved from timeless rock by the huge ice sheets of the Pleistocene ice ages, when sea levels were lower than they are today. Admiralty Bay may be a microcosm of all Antarctica. It is oceanic, it is terrestrial, and its heart is glaciated, but during the summer its shores are warm and rainy.

During the ever-bright Antarctic summer day, the sun marches around the northern horizon and only briefly dips from sight. The nights never really blacken but are long and often pastel twilights.[2] The summer bay is a huge nursery, denizened with all manner of life: humpback whales, elephant and leopard seals, giant petrels, skuas and terns. The Weddell seals spend the winter in the bay, chewing breathing holes in the pack ice. By October the females have climbed onto the ice to give birth, and for a few days the ice is stained red with natal blood. The penguins — first the Adélies and gentoos and then the chinstraps — arrive from September to December. The males stake out their little plots and shortly after are joined by the females. By the time I arrive in mid-November, it is already late spring. The big-eyed elephant seal pups are six weeks old and weaned. They lounge on the beaches, mustering the courage to venture into the sea for the first time. The Weddell seal pups, although still nursing, are not far behind.

In the mouth of the bay, which is five kilometers wide and subject to the passions of the open sea, penguins weave through the swells on their way to nesting colonies on the shore, looking like small piebald porpoises. The whole western entrance of the bay, from Demay Point to Point Thomas, is a guanoed penguin metropolis. During the summer, amorous penguins, each pair defending a modest cairn of pebbles, position themselves over the low hills and beaches with geometrical precision, exactly one pecking length in every direction from their neighbors. The rookeries are a cacophonous bustle of activity: bickerings, ecstatic displays, pebble robbing. By December the low volcanic hills are buffed pink with guano, and when the wind is westerly, the stench of the rookery wafts kilometers out to sea.

Behind the penguin rookery, the shoulders of the mountains rise in tiers of snow and black scree, capped with ever-present glacier. The Antarctic terns, black-headed and sharp-winged, lay their eggs in the shelter of these wind-roamed rocks. They course over the beach and sea, screaming at intruders, on their way to clip fish and krill from the bay. From prominent vantage points behind the penguin rookery, the skuas have set up their vigil for unguarded eggs and early chicks. They fly low and fast, just above the upturned heads of the nesting penguins, trying to evoke an inopportune lurch or other distraction to snatch away a hatchling and take it to the edge of the rookery, where two of them, tugging at the squealing baby, will tear it apart and eat it.

The bay has gradually warmed during the past century. Some of the glaciers have been reduced to small patches, isolated from the snowy interior of the island, and are evaporating, leaving in their slow wakes ablation moraines. Most of the ground surface is permafrost, but on warm summer days, earthflows of saturated soil and rock, upheaved by the cycle of freeze and thaw, ooze down the slopes. The stable areas of the mountain slopes are cloaked with fruticose, or branching, lichens. These are species of *Usnea,* and although only six centimeters tall, some individuals may be hundreds, perhaps thousands, of years old. *Usnea,* fretted and complex, look like basket sea stars at the bottom of a tropical ocean, and if it is snowing on the hillside — the big, moist, ephemeral snowflakes of summer — the flakes get caught in the tangle and are held, for a moment, until they melt. Relieved of the burden of winter snow, the lichens expand rapidly in the wet early-summer days. The new tissue of the growing edge of fungus is pale white-green, but the more mature spikes are dark green, and the very old, dead spikes are black.

Other parts of the bay have no beaches or mountain flanks but steep, cerulean glacial walls thirty meters tall. You risk your life taking a small boat near these facades, for they disintegrate without warning. The bay changes color according to the wind and the amount of wash from the glaciers. When the wind blows from the land, the water is rich in suspended sediments and is a pastel chalky green, but when the wind pushes choppy ocean water into the bay, the water is dark and clear. The ocean currents also bring swarms of krill, followed by the minke and humpback whales that graze on that pink bounty. Sometimes the humpbacks will loiter just offshore, indolently waving their white pectorals in the air and slapping the water.

The summer pulse of procreation lasts only a few months. By February the baby seals are independent, and the fledged penguin chicks, after long contemplation on the beach, are making their first forays into the sea. Fat and buoyant, they are easy prey for marauding leopard seals. Hard times set in for the skuas, which scavenge and bicker over scraps on the beaches, and by April the terns and the whales have migrated north. Once I lingered until late March and at last found darkness. Others have spent the winter at Admiralty Bay. Some winters the bay never freezes, but in most years snow begins to accumulate in March or April, and by late May the bay begins to glaze with ice, starting at the foot of the glaciers, where the water is freshest and coldest. At first the brisk glacial winds sweep the new ice out to sea, but by July the pack ice forms and the bay freezes solid. Now the day is a long twilight and the nights are black. The mountains that surround the bay become hoary with snow; only the tips of the volcanic nunataks and the sharp edges of the mountains remain brown.[3] The penguins follow the expanding edge of the pack ice far out into the Southern Ocean. The terns migrate to the warmer margins of Africa and the Americas. The skuas soon follow, although a few may stay the winter, especially near human habitations, where they can feed on garbage. During July, August, and September, the bay is dead and white.

* * *

Beyond the hilly penguin rookery at Point Thomas, across a green, mossy plain, is Arcktowski, the Polish research station. Sheltered in the rock at the point itself is a wooden Madonna who stares unblinking at the snow-dusted sea. Her face is sorrowful. The rocky beach is strewn with crates and oil drums. The station, behind a cockscomb of Polish, Russian, and Belgian flags, declaring the nationality of the scientists working there, is a series of rambling wooden buildings, standing on stilts above the permafrost, which intersect in a communal dining hall and kitchen. Once, sitting at the long dining table there, I sensed an unspoken tension between the Poles and the Russians. "We are all scientists and indifferent to politics here," a Polish biologist told me, between mouthfuls of *pierogi*. But he was wearing a *Solidarność* T-shirt.

Indeed, the primary purpose of every nation's Antarctic stations is political; the science (even good science) is just an excuse for a presence on the continent. Britain, Argentina, and Chile all claim the South Shetland Islands and the Antarctic Peninsula as their territories or dependencies. Regardless, since the 1960s Antarctica, with unknown but potentially vast resources, has become a free-for-all, and King George Island, only a three-day sail from South America and surrounded by a sea that is ice-free for much of the year, is the easiest place to set up shop. In the 1960s and 1970s Chile, Argentina, Great Britain, the Soviet Union, Poland, the United States, and Italy all established (and in some cases quickly abandoned) bases or refuges on the island. And in the first six years of the 1980s, Brazil, Uruguay, Peru, South Korea, and the People's Republic of China, all nations with negligible Antarctic heritage, also invaded King George Island. By necessity all of these facilities were built on the five percent of the island's coast that is not glaciated.

Parts of King George Island are rapidly becoming the urban slum of Antarctica. Some people, optimistically, consider it a sacrifice so that other areas may remain pristine. The worst hit is the Fildes Peninsula, on the southwest tip of the island, where once there were extensive meadows of lichens and important breeding colonies of penguins. It was one of the largest ice-free areas in all of Antarctica, and large areas of the peninsula were designated to be specially protected according to the terms of the Antarctic Treaty. But then came the Soviet invasion. In 1964, having failed to establish a station farther south on inclement Peter I Island, and with winter fast approaching, the Soviets hastily threw together a station on the Fildes Peninsula, smack in the middle of a specially protected area. (The next year the treaty was modified and, in true Soviet tradition, history was rewritten.) They named the station after Admiral Thaddeus Bellingshausen, the pioneering Russian explorer who sailed in these waters in 1821. It was not a happy place. During my visits to Bellingshausen I sensed a spirit of burdensome exile, of hierarchy and pecking order. The corridors were long and empty. The only perceptible scientific activity was at the weather station, which received and printed photos taken by Soviet and American satellites. In the recreation hall, a gloomy portrait of Lenin peered down at the pool table, as if he were calculating angles and shots.

Next to Bellingshausen, across a seasonal creek, are Base Frei, the Chilean meteorological station; Teniente Marsh, the Chilean air force base; and La Villa de las Estrellas (Village of the Stars), the Chilean Antarctic colony. Frei Base was constructed in 1965 during the Marxist government of Salvador Allende at the invitation of the Russians, who considered the Chileans to be comrades in socialism. But then the neighborhood changed, and the two stations have since endured a chilly proximity. The Fildes Peninsula was doomed. The Chileans built a long airstrip, able to accommodate four-engine C-130's, with adjacent hangars that sheltered a small air force of Twin Otters and helicopters. A hotel and cafeteria were built for the workers. Vast areas of the peninsula were graded for roads. And then came the colonists, hoisting the standard of Chilean nationalism. Their ranks include mothers and children, the families of the base officers, who sign up for two-year stints. They are Antarctic pioneers. The children, muffed in red parkas, bring the strange squeals of youth to the Antarctic. But they aren't exactly roughing it; the base has a bank, hotel, and gift shop, and a suburb of comfortable, rambling ranch houses with satellite TV. One evening that I spent at the hotel, the guests gathered in the salon to watch "Miami Vice."

✳   ✳   ✳

I first sailed into Admiralty Bay on a sun-swept day in late November 1982 and made a landing deep in the interior of the bay at Martel Inlet, on the shore of the Keller Peninsula. Across the inlet was the Ternyck Needle, a nunatak of brown basalt that punched a hole in the glacier and rose 90 meters above the ice cap. The wind had scooped a trench 25 meters deep and 45 meters wide in front of the nunatak, which seemed incongruous and disturbingly out of place, an alien rock in the field of ice. Keller Peninsula was a spine of extinct, eroded volcanos, 600 meters high, that emerged from the glaciated interior of the island. To the east was Stenhouse Glacier, which originated in the ice dome of the island's interior and slid down a valley of its own making to the sea. Like all of the glaciers on King George Island, it was shrinking, calving icebergs and evaporating faster than the snow could accumulate on the ice dome above. Years before, British glaciologist G. Hattersley-Smith journeyed on dog sleds and skis over this glacier to measure its rate of travel. He found that it varied during the summer from 23 centimeters per day high up on its heavily crevassed slopes to over 100 centimeters per day where the ice is warmed and softened by the summer sea. During the winter, when Martel Inlet is shaded by the spine of the island and the sea freezes, the glacier stops calving and rests.

Behind the beach was Flagstaff Mountain, an extinct volcano 600 meters high, and on its flank a vestigial glacier of the same name. Flagstaff Glacier was only a few hundred meters across and was not joined to the ice cap in the interior of the island, nor to the sea. With no new mass flowing into it, its only crevasse choked with snow, it was slowly evaporating. On this summer afternoon it was dusted with pink snow algae.

I found two derelict huts on Keller Peninsula: Argentine and British meteorological stations. Both were built in 1947 during a squabble over Antarctic territories. The Argentine station was abandoned in 1954, but the British station, which bore the uninspiring name of Base G, remained occupied, summer and winter, until 1962. The boards used to construct the base were scavenged from the old whaling station at Deception Island. The grain of the wooden planks was so etched by wind-blown ice and dust that every detail of the knots and whorls stood out. The hut's windows and doors were stove in and its rooms were filled with snow. Rotten wooden tables and chairs, rusting bed springs, canned food with peeling labels, molten paperbacks, dissected motor parts, and tools were all frozen in their places, just as in the moment of abandonment twenty years before. On the floor of the hut was a dilapidated plaque that read:

> The preservation, care and maintenance of
> these historic ruins has been undertaken
> hereinafter upon the aforsaid [sic] schedule
> by the Ministry of Perks [sic].
> Permission to view from
> the Chief Magistrate

Behind the stations on the first ridge of rocks that overlooked the pastel bay were two freshwater ponds fed by melting snow. I peered into one of them. The water was dancing with copepods. Above the ponds four lonely crosses, memorials to casualties of the British station, were etched silently against the horizon. They coldly endowed the peninsula with a feeling of being peopled; one never felt quite alone there. I hiked up the ridge to read their inscriptions.[4] On the first cross was written:

> IN MEMORY OF
> DENNIS RONALD BELL
> BORN 15·7·34
> WHO WHILST SERVING WITH F.I.D.S.
> DIED ACCIDENTALLY AT BASE 'G' ON
> 26·7·59

The second cross:

> ERIC PLATT
> GEOLOGIST
> BASE LEADER
> F.I.D.S.
> DIED ON DUTY
> 10 XI 1948
> AGED 22 YEARS
> R.I.P.

On the third:

> IN MEMORY OF
> ALAN SHARMAN
> BORN 29·12·36
> WHO, WHILST SERVING WITH F.I.D.S.
> DIED ACCIDENTALLY AT BASE "G" ON
> 23·4·59

I realized that these explorers, who had died on the rim of the earth, were just boys. This nursery bay, lovely on that still afternoon, could take life as well as bestow it.

The fourth cross was so eroded that it read only:

> R   G.   N
>   B   L
>   F.I.D.S.
>   B
>   L

The hero's name had been lost to the wind, erased by blowing needles of sand and ice.[5]

The pebble beach in front of the huts was littered with stranded icebergs, sculpted into fantastic shapes by the spring thaw, and a covered wooden boat that had been used to haul water a half-century earlier by whalers who took shelter in the bay. Its iron chains were decomposed and bleeding rust, but the wooden rivets that bound the hull were still preserved. What I remember most, though, were the hundreds of whale bones — skulls, ribs, and vertebrae, looking like a giant child's scattered jacks on the beach. They were far more permanent than any human-made structure. To the north, beyond the Argentine station, the twenty-meter-long skeleton of a baleen whale, a composite of many individuals of several species, had been reconstructed on a bed of moss behind the beach by Jacques Cousteau and the crew of the *Calypso,* which occupied Base G during the summer of 1972–73. The whale's deflated ribs were splayed on the ground, its vertebrae looking like a white picket fence. It was late spring, and a shelf of melting snow and ice covered the beach below the whale. The silence was absolute, except when an iceberg calved from Stenhouse Glacier with a reverberation that ricocheted across the bay and the white snowbanks of Ullman Spur. This was followed moments later by a small tsunami, which surged onto the beach and shuffled the icebergs.

* * *

I next visited the bay in 1987. By then the Brazilian research station had popped up like a cluster of metal and plastic mushrooms on the beach, behind the highest tide mark, between the Argentine hut and Base G. I had been invited to the Brazilian station for the summer to conduct marine biological research. Events in Brasília, 4,800 kilometers to the north, had motivated the politicians and generals to declare that tropical Brazil had a national interest in Antarctica. An illuminated crucifix had been hauled to the top of Flagstaff Mountain, transforming it into a miniature Corcovado. A pumphouse had been built next to one of the freshwater lakes to provide the station with water, carefully filtered of copepods (at the other end of the cycle, the sewage was pumped into a deep well). The station consisted of fifty-one insulated tractor-trailer containers, painted pale green, linked by narrow corridors much like railroad cars, giving it a boxy, unfinished appearance. The green and yellow Brazilian flag, bearing a globe on which was inscribed "Ordem e Progresso" (Order and Progress), shuffled in the wind from a mast on the beach. Behind it was a soccer field.

Huge rubber bladders of fuel wallowed on the beach, oozing oil into the bay. The industrial age had come to Admiralty Bay. The tranquility that I so vividly remembered from five years before was gone. A diesel generator, which ran twenty-four hours a day, shredded the silence, and I had to walk beyond the old Argentine station, 200 meters south, to be free of the noise. The generator was the metabolic heart of the station, an artificial sun that kept the personnel warm, incinerated their garbage, and provided the power to link them with the outside world.

"Polar exploration is at once the cleanest and most isolated way of having a bad time which has been devised," wrote Apsley Cherry-Garrard more than a half-century ago.[6] But we had fun. Visitors affectionately nicknamed the Brazilian station "Little Copacabana." Its spirit was buoyant and welcoming. Just inside the door, one stripped off boots and woolen clothing in an arid, hot foyer and hung one's clothing on the wall. The walls were lined with hooks, each designating someone's little vertical territory: a pair of boots, a red parka, gloves and a hat. Three hooks were reserved for visitors. As in any Brazilian household, the kitchen was the center of social activity. Coconut snacks and sugary *cafezinho*, served in demitasses, were always available on a wide counter covered by a plastic tablecloth printed with Santa Clauses. The automatic dough-kneading machine plodded like a metronome, filling the air with yeast smells. A traditional Saturday feast of *feijoada completa*, a Cariocan dish of black beans, sausage, and pig's ears, was simmering on the stove. Behind the kitchen was the radio room, where a squawking box linked the station with the disembodied voices of lovers, children, and bureaucrats in the crowded parts of the world. In the central dining room and lounge, on a cluster of sofas, a group of scientists watched a video of *Back to the Future*, and in another corner, someone was playing a kung-fu game on a Brazilian clone of the Apple II computer. The electronic imitation of karate kicks became a constant acoustical backdrop at the station during my visit. The walls were paved with the emblems of ships that had visited the station, the banners of Rotary and Lions clubs, portraits of station members over the years. On the western wall was an oil portrait of Comandante Ferraz, one of the first Brazilians to travel to Antarctica, after whom the station was named. In the far corner was the bar, which also served as a post office (you canceled your own letters and they were picked up every two months). It was stacked neatly with crystal brandy snifters to be used only for special events, which, because of the Brazilian *joie de vivre,* occurred often. The bar also stored the accordion, drums, and tambourines. Parties often lasted all night, and *Carnival* lasted three days. Behind the bar was mounted a fossil leaf of a southern beech tree, 16 million years old, found at Cape Hennequin across the bay. One eastern window of the lounge had been shattered by a tempest-flung rock, and the shards had been colored with cellophane and rearranged in the form of a *papagayo,* the cloth and bamboo parrot kite of Copacabana Beach.

Beyond the lounge was a warren of sleeping modules, each equipped with a set of bunk beds, a small table, and a double-paned window. A central module contained the toilets and showers, men's on the north, women's on the south. The water was heated to scalding by the exhaust of the generator. Only the commander, Queiroz, who by tradition was a navy captain, had a private module and bath. Next to his module was the gym, the only structure that was not modular, with tall windows overlooking the expanse of black beach. It was cluttered with mountaineering equipment, skis, barbells, and weights. At its center was a snooker table, plenty of chalk and cues, and a slate blackboard with a system of numbered beads for keeping score. The clack of billiard balls was another pervading sound of the station.

The hospital, aquarium, and biology laboratories were separate from the living modules, under a gloomy metal roof that protected them from drifting snow. The base physician, deputy commander, and postal clerk was Dutra, an urbane doctor from Rio de Janeiro, who kept the operating room scrubbed and instrument-ready for sudden emergencies.

I spent much of each day, and the long bright night, in the three biology modules. We had to wear warm parkas while working in the aquarium module, which was kept at near-freezing in order not to poach the heat-sensitive organisms that we were studying. On some nights, while manipulating the cold steel knobs of the microscope, I would wear gloves. On very windy days the module would shake so much that it was impossible to use the microscope. It was a damp, cold, noisy place, filled with the gurgle of filters and the hiss of water jets, which were continually charged with seawater pumped through a black PVC pipe that snaked over the rock beach from the bay. The water supply had to be constantly monitored to keep it from freezing. If the catch had been good that week, the nets and baited traps not snarled by drifting ice or, worse yet, lost, the aquariums were populated with all manner of marine organisms from the bay: brown antifreeze fish, big-mouthed, nacreous icefish, starfish, sea anemones, isopods, amphipods, sea spiders, and krill. A few of the krill were forced to swim against an artificial current in a physiological chamber, their oxygen consumption measured by instruments that periodically tasted the water. This corner belonged to Phan, a Vietnamese-Brazilian physiologist. The Brazilian station lacked the trawl nets and swift boats necessary to capture krill on the open sea. But Phan had devised a system of high-powered lights that attracted krill (and myriad other organisms) close to shore in front of the station, where they could be conveniently caught in hand nets. At the opposite corner of the laboratory were the glass aquariums used for behavioral studies. This was Claude's corner. He spent long hours staring into the aquariums to observe the feeding strategies and food preferences of captive amphipods, trying to imagine what they would be doing on the unseen bottom of the bay.

The two other labs were under a sheltering canopy and therefore slightly warmer, but you still needed to wear a heavy coat inside. These contained more microscopes, dissecting tables, cupboards of chemicals, scales, plates of agar, and sinks with hot and cold running fresh water. On the days the nets were brought in, the labs reeked of formaldehyde. Renato, the veterinarian, worked here. It was Renato who had invited me to work at the station as part of a team trying to figure out the life cycles of parasites that live in seals, fish, and crustaceans.

During the summer Little Copacabana was a favorite stop for the tourist ships, which would disgorge legions of passengers, all clad in red parkas so that their leaders could keep track of them. Most visits lasted only an hour or two, and the tourists didn't learn, or see, much. After the obligatory pose for a portrait with Cousteau's whale skeleton, they shuffled into the station to buy stamps, postcards, key rings, and other mementos utterly irrelevant to Antarctica. Some offered to barter T-shirts and other wampum with the station personnel, as if they were primitive tribesmen. Snuffling and sneezing, the tourists usually left colds in their wake. Regardless, the hospitable Brazilians had a limitless tolerance for these intrusions. Often they would organize a soccer game with the ship's crew and break out the *cachaça* and the accordion.

The tourists were innocuous and friendly, but the visitors who made us truly uneasy were those from Greenpeace, which had a ship patrolling the area. Ready with criticism and armed with video cameras, Greenpeace has brought to world attention many environmental problems in Antarctic stations and has advocated the concept of an Antarctic world park. "If they come, don't offer them any information," cautioned Queiroz. "Let them ask first." When Greenpeace visited Comandante Ferraz in April 1987, the Brazilians, true to their nature, threw them a party. Greenpeace reported that "the Brazilian station is one of the tidiest seen. . . . It is obvious that they not only took good care of their base, but also the surrounding environment."[7]

The station had a homey ambience the summer I was there. But we were probably just lucky. Crews for Antarctic stations are like classrooms full of students: each has its own, unpredictable collective personality. But the psychological stress of prolonged confinement, with minimal privacy, in modules crowded in a hostile environment, often gets to even the strongest individuals. Antarctic workers frequently develop a dazed, almost autistic condition known as "big eye." Therefore the screening of candidates for the stations is serious business, particularly for the personnel who will endure the long, dark winter. At one of the Soviet Antarctic stations a worker killed another with an axe in a dispute over a chess game. During the autumn of 1983, the staff doctor at the Argentine station Almirante Brown, on the edge of Paradise Harbor, 400 kilometers south of King George Island, forced his own rescue by burning the base down.

Candidates for the Brazilian Antarctic program were evaluated according to fifteen criteria, including "resistance to long periods of confinement, resistance to frustration, capacity for prolonged sexual abstinence and capacity to sustain long separation from one's family." How the evaluator could possibly ascertain this information was beyond me. In addition to the eight support personnel from the navy, all men, Ferraz Base had twenty-one scientists, five of whom were women.[8] In many ways the station personnel become a surrogate family, and, like all families, they may be supportive or abusive. The Brazilians were wonderfully supportive, and we hugged and cried at the end of the summer in one *grande tristeza* of homesickness that lingers even now.

*   *   *

This book is about living for a summer in that Antarctic outpost. It was sometimes easy to forget, in that snug station on the edge of a blue bay, that we were living in an alien environment, beyond the edge of the habitable earth. Only the generator and a few membranes of metal and cloth prevented us from freezing to death. But the station, although necessary for our survival in Antarctica, was an upstart intruder in a timeless, frozen place. We were scientists who had to come to study more enduring things: fossils and glaciers, the ebb and flow of seasons, wind and albatrosses, metropolises of penguins, and the crowded, unseen Antarctic underwater realm, which brims with life as no other sea on Earth. We were pilgrims in the last new land on Earth.

1
Seabirds and Wind

An albatross, whose wide-spread wings asleep
Lazily follow, poised above the wake,
The vessel gliding o'er the bitter deep.

. . . Thus with the Poet! though he ride the cloud,
Soar with the storm in skyey wanderings,
Exiled to earth amid the howling crowd,
He stumbles, shackled by his giant wings.

— CHARLES BAUDELAIRE

GETTING TO ANTARCTICA, whether by airplane or by ship, is often uncomfortable and always dangerous. Tonight, the beginning of my third summer in the Antarctic, I board the *Barão de Teffé*, the supply vessel of the Brazilian navy, at Punta Arenas in Chilean Patagonia. At the hour of our departure the rotund Brazilian consul is at the quay with a coterie of women and children, who hold hands in a large circle and sing seafarers' songs of loneliness and isolation. A lone woman in the front seat of a green Subaru seems to be sobbing and praying. The sailors are morose. There is a certain unease tonight. What does the Drake Passage have in store for us? Crossing the roaring forties and screaming fifties by ship from Patagonia can be serene, on wide, slow swells. Or it can be a nightmare, on seas so madly kinetic that you have to brachiate from handrail to handrail and adapt to knee-buckling extra G's as the ship rises, and near weightlessness when she falls. If you stand on the poop deck, you hear the propeller turning in the air one second and are inundated by freezing sea the next. And even if you don't become seasick, the constant battle against the ever-shifting sea dilates every mundane activity and makes you grumpy and tired.

Last week the *Profesor Besnard*, the research vessel of the University of São Paulo, sheared her drive shaft in the Drake Passage and, without control in the high seas, almost foundered. The captain radioed an SOS and the crew was ordered to the lifeboat stations, although launching the open boats in the high seas would have been suicidal. The metal tables in the laboratory, located over the bucking stern, were bowed by the extra strain placed on them during the wave shocks. But the *Profesor Besnard* was lucky. As abruptly as the storm started, it marched eastward, and the ship was safely towed to port.

The *Barão de Teffé* was built in Denmark specifically for expeditions to polar regions. She has changed hands several times, first serving the Danish Lauritzen Line (the blankets and towels still bear the insignia of that company) and later carrying the French flag before being purchased by Brazil. Her charter Brazilian captain was Comandante Ferraz, who died while bringing her to Brazil from Europe. Her ice-strengthened hull is painted bright red for quick identification, and if need be, rescue, among the white ice floes.

She is a clamorous vessel. The hydraulic steering mechanism, located behind my cabin, whines and clinks every few seconds. Each cabin has a ghetto blaster playing Bronx rap music, which is all the rage among the crew; only rarely does one hear an indigenous *samba* or *tropical*. The ship is overcrowded, and three or four people are crammed into cabins made for two. My two roommates are engineers who preen the diesels. Their orange deck clothing, grimy with grease, hangs from hooks on the door. Career officers, they alternate five months at sea with seven at home in Rio de Janeiro. They have developed the easy amiability that is requisite for survival in crowded quarters for long periods at sea. But still they have become slightly stir-crazy, spending their leisure hours playing cards and dominoes, slapping the chips on the Formica table with reverberating cracks, and sipping *maté* from ornate *chimarrãoes* (gourds) through silver straws named *bombas*. In southern Brazil the passing of the *chimarrão* is a gesture of friendship, like the passing of a peace pipe; I was offered a *chimarrão* as soon as I entered the cabin. We live in a claustrophobic world stripped of privacy, a realm of competing stereos, humid bath smells, a pall of cigarette smoke. You have to close the lid on the head before pumping the flushing mechanism, else it burps in your face. The grubby brown pile carpet is constantly damp with burp water. I sleep, wrapped in wool blankets, on a sofa that is impregnated with engine oil and cigarette ashes. It is at water level just under the porthole, which turns sea-green and then sky-bright every few seconds. An ice-cold condensate drips from the brass clamp of the porthole; I cover my head with the wool blanket and try to ignore it.

The journey takes four days. Our days and nights are regulated by a series of whistles and bells marking the watches. This an anachronism from the days when ships had no telephones. At his appointed hours, the officer on watch walks purposefully through the lurching corridors, blowing his whistle like a child on the loose with a toy. The whistles cause every conversation to pause, wake every sleeper, but they also convey the comforting news that all is well. You can roll over and sleep a little easier.

The scientists eat in the officers' mess. The beer-bellied steward is always dressed in a T-shirt with anchors and eagles painted on it. There is a fixed pecking order at the table. Before lifting a fork or sipping our coffee, we must wait for the first mate to begin eating. Sometimes he doesn't appear, and the steward must phone the bridge for permission to start. Then he walks from table to table giving the thumbs-up signal, as if blessing each hungry group of sailors. The dining room table has a curb around its edge and rubber pads to keep the dishes from spilling onto the floor

during a rolling sea. For breakfast, at seven bells, we eat Sucril-
hos, sugar-frosted flakes, with hot milk. *Antonio o Tigre* grins
from the box. The coffee is delicious — dark-roasted and viscous
with sugar — but most of its caffeine has been burned out and it
doesn't pack any zing. Lunch is liverwurst and lots of beans;
dinner, meat and lots more beans.

   The social center of the *Barão* is the *praça das armas,* which in
the old days of sailing ships was the armory. On the *Barão* it is a
lounge and bar. It has a television, tables heaped with four-
month-old magazines, chess, and checkers. I have renamed it
O Salão Grande do Povo (the Great Hall of the People), and this
seems to amuse the scientists, although it is considered vaguely
disrespectful by the officers. There is a party in the *praça das
armas* tonight, celebrated with slugs of duty-free Stolichnaya and
various friendly toasts. The seas are too rough for dancing. The
music is Broadway, Engelbert Humperdinck, and reggae — the
shards of alien cultures. I am asked to translate the lyrics, which
are cryptic even in English, and incomprehensible in Portuguese.
How do you explain "Every Little 'Ting Gonna Be Alright" and
"Follow the Yellow Brick Road"?

*   *   *

Antarctica begins not at the edge of the pack ice, or in the fastness
of glaciers, but invisibly, at sea, where the cold polar surface
water slides beneath the slightly warmer water from the north.
This is the Antarctic Convergence, an undulating front of water
masses that completely circles the Antarctic continent. Today, at
dawn, halfway between Cape Horn and the Antarctic Peninsula,
we crossed this subtle barrier. One does not at first notice the
transition. Over a stretch of about 80 kilometers the water temperature drops a few degrees to near zero centigrade, bringing a
gray-blue haze and a penetrating clamminess. The southern horizon is lighter, and beyond it is unseen pack ice; early polar
mariners named this phenomenon "ice blink" and used it to avoid
the pack ice. Soon the first floating ice begins to appear: a wind-

blown litter that rattles and grinds down the length of the ship, as well as flat-topped tabular icebergs, sixty meters tall, which look like phantom continents. In this sector of Antarctica, tabular icebergs break from the 300-meter-thick Larsen, Ronne, and Filchner ice shelves of the Weddell Sea. Some are kilometers long and generate their own local weather of gauzing fogs and katabatic winds.[1]

In 1898 the American polar explorer Frederick Cook described his passage over the Convergence, somewhere near this spot, in the bark *Belgica*. In those days little was known of the physical characteristics of the boundary. Cook knew only that far out to sea was the beginning of Antarctica. "The night which followed was dark," he wrote.

> The sea rolled under our stern in huge inky mountains, while the wind scraped the deck with an icy edge. We kept a sharp lookout for icebergs, which might come suddenly into our path out of the impenetrable darkness ahead. The sudden fall of the temperature and the stinging, penetrating character of the wind seemed to warn us that ice was near; but we encountered none. Life was plentiful, but melancholy. Curious albatrosses and petrels hovered about us, uttering strange cries, and in the water there was an occasional spout from a whale. It was a night of uncertainty, of anticipation, of discomfort — an experience which only those who have gone through the wilderness of an unknown sea can understand.[2]

To the traveler sailing south toward Antarctica, the Convergence is the first hint of things to come and of the complex layering of the Southern Ocean. Every spring the Antarctic surface water is diluted by the meltwater from the pack ice and the glacier-born icebergs that calve from the continent. Near freezing and low in salinity, the surface water spreads slowly north. But no water can be exported north from the Southern Ocean unless it is compensated for by other water masses heading south. This balance is maintained by the warm deep water that flows, 50 to

100 meters below the surface, from the temperate Atlantic, Indian, and Pacific oceans, under the Convergence to the edge of the Antarctic continental shelf. There it upwells, spewing nutrients onto the shelf and, each spring, fostering a bloom of phytoplankton. This upwelling is known as the Antarctic Divergence and, like the Convergence, is a zone splendidly rich in life. Part of the Divergence heads north, eventually descending at the Convergence and creating a huge vertical gyre; part heads south, right to the edge of the continent.

The water that heads south from the Divergence collides with the continent, eventually sinks, and joins the deepest layer of all, the north-flowing Antarctic bottom water. Like the Antarctic surface water, the bottom water is born of ice. But it is created when ice is made, not melted, and is therefore opposite in character from the light, low-salinity surface water. When sea ice is formed, the salt is concentrated in channels of brine, which slowly work their way to the bottom of the floe and are extruded into the sea. The water under the floe is therefore near freezing and high in salinity, both conditions that make it dense. The Antarctic bottom water slithers far north on the belly of the sea, over the edge of the continental shelf and into the abyssal zones at the bottom of the oceans. It oozes beneath the tropical seas, crosses the equator and insinuates itself into the deepest areas of the North Atlantic and North Pacific, a journey that takes decades, perhaps hundreds of years. All the while, the temperature of the deep water remains near freezing. In the Bahamas, for example, the three-kilometer-deep Tongue of the Ocean, between Andros and the Exuma cays, is paved with frigid Antarctic bottom water. The abyssal areas of the world's oceans are therefore remarkably uniform: born in Antarctica, dark, salty, cold, and subjected to enormous pressures. And the life forms found there are also remarkably similar: giant isopods, glass sponges, soft corals, and slow-moving fish with low metabolic rates. Many of these organisms are adapted to feed on the gentle rain of organic material from the surface. It is no coincidence that these animals also occur in the shallow seas around Antarctica. A scuba diver in the Antarctic can observe types of animals that elsewhere on Earth can be seen only from the portholes of deep-diving submersibles.

The Antarctic Divergence is not only a frontier of the vertical separation of water masses, it also marks the lateral shearing of ocean currents and winds. South of the Divergence and close to the continent, a narrow band of currents plies westward, pushed by the cold easterly winds that spin off Antarctica. This is known as the East Wind Drift by the few sailors who have ventured so far south. North of the Divergence, the opposite takes place. A massive band of the Southern Ocean, from nine hundred to several thousand kilometers wide and extending beyond the Convergence, is pushed inexorably east by westerly winds. This is known as the West Wind Drift and is the zone of the roaring forties and screaming fifties. The incessant winds that generate the West Wind Drift completely isolate Antarctica from the other continents. Antarctica is alone at the bottom of the world amid a swirl of sea and storm.

The Antarctic Convergence is perhaps the longest and most important biological barrier on Earth, as formidable as any mountain range or desert. It is an obstacle to the dispersion of birds, fish, and, most important, plankton. South of the Convergence the seafloor is smothered with a soft glassy ooze of diatom shells, the accumulation of a near infinity of one-celled plant bodies that have rained down from the sunlit surface waters. But the warmer waters north of the Convergence are dominated by another phylum of one-celled algae, the coccolithophorids, and the seafloor is covered with their calcium skeletons. Both diatoms and coccolithophorids are invisible to the naked eye, yet the shift in their relative abundance affects all of the species that directly or indirectly depend on them for food. Diatom-eating Antarctic krill range only as far as the Convergence, for they would perish in the stifling warmth to the north.[3] The species of squids are also different on either side of the Convergence.

South of the Convergence there are only thirty-nine species of nesting birds — seven penguins, six albatrosses, eighteen petrels, one cormorant, two skuas, a gull, a tern, a sheathbill, a pintail, and a pipit — fewer species than can be found in a small garden in Colombia or Costa Rica, but numbering about 70 million individuals.[4] All but the pipit and the sheathbill are seabirds, which consume about 7.8 million tons of krill and other zooplankton per year. In spite of their low diversity, the density of birds is often astounding. On South Georgia Island more than a hundred nesting burrows of white-chinned and blue petrels have been counted per hundred square meters, creating huge avian metropolises that ramble for kilometers over the lumpy meadows of tussock grass.[5] Volcanic Zavadovskiy Island, in the South Sandwich archipelago, has a colony of approximately 3.5 million chinstrap penguins. This is the recurring evolutionary motif in polar areas: a paucity of species but an abundance of individuals.

* * *

Today, halfway from Patagonia to Antarctica and just south of the Convergence, I watch seabirds from the bucking prow of the *Barão de Teffé* and muse on the complex interactions of water and air. The bow cuts the Antarctic surface water; sixty meters below me is the warm deep water; a kilometer and a half or so below me is the lightless and practically unknown realm of the bottom water.

A ship sailing through the roaring forties toward Antarctica is always trailed by a coterie of seabirds. Almost all are broadly classified as tube-nosed birds, a reference to the salt-excreting glands on their bills. The tube-nosed birds include the petrels, prions, albatrosses, fulmars, and the aptly named shearwaters.[6] They dust the sea for days and nights on end, like gnats on a stormy lake. In the chaos of wave and bird, and from the pitching deck of a ship, it is often difficult to tell one species from another unless you are close enough to observe the way they feed and the way they fly. These behaviors also reveal the separate niches of these birds, their ways of avoiding direct competition with each other.

The giant petrels, which with a wingspan of 2.4 meters are the largest of the petrels, glide on locked wings, always tilted, so that one wing tip appears to softly touch the sea surface, then they soar, again always tilted, and dive again. The Antarctic prions cut furrows in the water with their heads; they have broad bills with laminated palates (like those in a flamingo's beak) that filter plankton from the water. The narrow-billed prions fly slowly, tipping like stiff-winged kites over the sea surface, and pick individual plankton from the first few centimeters of water. If there is an abundance of plankton, or a sheen of mucus and oil from the exhalation of a whale, the prions will abandon wing and alight on the sea to feed.[7] Rafts of these birds hint of bounty below. The South Georgia diving petrels, small and pudgy, fly with rapid wing beats and short glides. Sometimes they fly straight through the wave tops, in and out of the crest before it collapses. But more characteristically they plunge into the sea in pursuit of individual krill or other large plankters. The pure white snow petrels are so perfectly adapted to their environment that they fade into white snow and sea spray. They fly with fast, urgent strokes, actually palpating the undulating sea surface with their sharp wing tips, feeling the shifting swells and waves. The southern fulmars are gray like cloud shadows and fly with pulses of wing beats interrupted by long, lazy glides. When they alight to feed, they stall in the wind with outstretched wings and drop, folded feet first, into the sea. The pintado petrels patter the water with webbed feet. Their backs are the color of sea foam on dark water; their bellies are pale as sky. This is their camouflage: each bears an inverted sea, sky, and horizon on its body in a realm that is only sea, sky, and horizon.

But the birds that astound me are the smallest: the Wilson's storm petrels. Only eighteen centimeters long, with a wingspan a little over twice as wide, they may be the most abundant bird on Earth, occurring in vast but unknown numbers in the Antarctic, the Arctic, and all oceans in between. Theirs is an erratic flight, with many changes of course. They flit among the waves like frail black butterflies, their white rumps and white wing chevrons clearly visible from above. Yet these little birds are almost impossible to track in the tumult of the waves. That they are able to fly over the complex, randomly shifting sea surface, these moving mountains of water fretted with reverberating pulses and shocks, seems to me the epitome of evolutionary eloquence. What cerebral wiring — and what programs — they must possess in that pea of a brain, safe behind hard skull and soft black feathers, to enable them to accomplish these shuttlecock navigations, these magical weavings in the random sea. Their metabolic fires have to be constantly stoked with pteropods, crustaceans, and the occasional small squid to power such energetic flight. In a sense, they are like the tropical hummingbirds, which flit among the blooms, unable to stop lest they burn out. But instead of sipping nectar on the wing, the storm petrels foot-patter the sea without alighting and, their wings almost vertical, lean forward to peck morsels from the water.

* * *

Antarctica is the windiest place on earth. Wind dominates all activity, sets schedules, imperils lives, makes one a hostage indoors. It even determines what species have colonized the continent. And it is wind, not earth or sea, that is the principal milieu of the albatrosses, which in the Drake Passage are the wandering, sooty, light-mantled sooty, and black-browed. They are aerial beings, who spend most of their lives on the wing above the undulating sea, except when they briefly alight to snag food or when they nest on shore. For hours on end I watch them from the poop deck. They glide a few feet away at eye level, adjusting their dip and speed to the velocity of the ship. What a privilege it is to share this space of air above the Southern Ocean with these creatures of the wind. In the sharp, clear light of the lingering sunset, I can discern the detailed fretting of every pinna (the primaries flex in the gusts), the razor-sharp bill (shocking pink in the wandering albatross), their roving eyes searching sea and ship. In these latitudes, the most abundant are the black-browed albatrosses, subantarctic birds that breed in the Falkland Islands and prefer the productive waters of the Convergence. They will follow us as far as the pack ice, which dampens the sea wind and robs them of flight. At first they seem as stiff-winged as a child's kite, but in fact their wings are wonderfully sensitive and dextrous. The wing *feels* the subtle eddies of wind along its length. Each wing bone and feather has a suite of muscles that can slightly shift its orientation and, in unison, adjust the wing's total geometry — the angle at which it is swept back, the degree at which it cups the wind, the total surface area — from second to second. Wired into the brain of the albatross are programs for physics, aerodynamics, meteorology, and global navigation. These animals instinctively *know* that wing beats are uneconomical and inefficient. They *know* that the velocity of the wind is lower near the water surface because of friction than it is fifteen or twenty meters above, and that by dipping from the brisk winds at altitude into the viscous surface winds and then rising by flying at right angles to the wind direction, they can steal the wind's energy and soar without wing beats.

The wandering albatrosses, with a wingspan of 3.3 meters, are the largest of all seabirds. They seem to be objects of sky itself, weightless beings that embrace the wind and float on it with a relaxed elegance. They are paradigms of economy and aerodynamic efficiency. The ratio of wingspan to wing width of a wandering albatross is eighteen to one, about the same as the most sophisticated human-made gliders. The wing's lift-to-drag ratio — of the lifting force to wind resistance — is forty to one. They are extremely light; their bones like those of most flying birds, are partially air-filled. The entire skeleton of a wandering albatross, including its skull, bulky sternum (to which the flying muscles are anchored), and wing bones, weighs only 1.3 kilograms. Wandering albatrosses *need* wind. Indeed, a flying bird of this great size probably could have evolved *only* in the wind-roamed Southern Ocean. A wind speed of less than 19 kilometers per hour forces them to beat their wings, and the halcyon calm of a high-pressure zone can strand them on the sea surface, unable to lift off, for up to a week. But when there is wind, their flying is prodigious. A radio-tagged wandering albatross that was tracked by satellite from its nest at Crozet, a subantarctic island in the Indian Ocean, flew more than 15,000 kilometers on a single foraging excursion of thirty-three days. At night, from an altitude of a hundred meters above the sea, they may be able to see the bioluminescent schools of squid that rise to the surface to feed, and alight on them, themselves to feed. When they are not grounded by the mundane activities of the nest, wandering albatrosses may repeatedly circumnavigate the planet. They are global citizens, crossing with indifference all of the crazy, invisible boundaries of humans. My companion, a few meters away, inscrutably watching me, is able to encircle the entire planet with its body, using nothing more than feathers, flesh, and bone.

Yet on the subantarctic islands where the wandering albatrosses breed, they seem to be in slow motion. They laboriously build wattle nest mounds, half a meter high, among the clumps of tussock grass and lay a single large egg. The incubation of the eggs takes nearly two months, and the chick's period of parental dependency is another nine to twelve months. This is due to the relative scarcity of food for such a large bird; there is economy — and protection from hard times — in slow growth. Timing is also important. Albatrosses lay their eggs from late December to early January, which ensures that the chicks will fledge at the peak of the summer zooplankton, and the squid that it supports, a year later.

One of the better-studied breeding areas of the wandering albatross is Bird Island, a low key a few kilometers north of South Georgia. I once spent a day on Bird Island. The ocean banks surrounding it were snarled with kelp, and I had to hang from the bow of the landing craft and cut a path with a machete. Once on the beach, we had to run a gauntlet of aggressive bull fur seals, which were defending harems of demure females and their pups.

Some of the bulls were perched on clumps of tussock grass, snouts pointed upward, as if imperiously surveying their domains below.

The wandering albatrosses nest on the wind-tousled hills over-looking the beach, well above the commotion of the seals. I visited Bird Island during early December, when the chicks were almost as large as their parents. Waddling over the grassy moors, they occasionally tested their long wings in the wind, but seemed afraid to commit. There are no albatross predators on Bird Island, and the birds are naturally tame. When I approached carefully, they allowed me to sit quietly next to them. But when I shifted my weight or moved too abruptly, they clacked their pink bills in gentle warning.

A decade ago, if an albatross chick survived its first year at sea, it had a good chance of living up to fifty years. But today an albatross faces many risks. Worldwide, there is a one percent decline in the numbers of wandering albatrosses each year; in less than a century — two generations — there may be few left. Al-batrosses must compete for food with the expanding flotilla of squid boats in the Southern Ocean, and many die in fishing nets. Some become so dependent on ships' garbage that they lose their hunter's edge and starve when the boats are absent. But as I squatted next to this great bird, this living evocation of the south-ern wind, I couldn't imagine that it might someday turn into a garbage addict.

✳   ✳   ✳

The Antarctic. The *Barão de Teffé* is anchored in a half-moon bay, sheltered from the north wind by the ramparts of Elephant Island, the northernmost of the South Shetlands, 200 kilometers from Admiralty Bay. The island broods off our prow, smothered in ice clouds. Sometimes the clouds part and the black-brown, sharp-edged basalt peaks appear. Between them are crumpled glaciers flowing from the mountains to the sea, their cracks and fissures a cold aquamarine. To the west is Lookout Point, its flanks stained russet by penguin guano — a splotch of krill on the mountain slope. To the east, Endurance Glacier seems wrought from light, blue and white.

Helicopters are clattering to and from the island, bringing supplies slung in hanging nets that resemble oriole nests for the ground team that will camp here and study nesting giant petrels. Rice, beans, fuel, blankets — all the peculiar, alien items needed to sustain humans in this hostile environment.

The sea in the bay is turquoise, reflecting the shallow bottom. An iceberg is grounded in the bay. The blue wash of the ocean smears on its smooth flanks or bashes on its steep ice cliffs, making fresh wounds. Its turreted top was no doubt once its bottom, the iceberg having turned over as the equilibrium of its mass shifted, melt drop by melt drop, calorie by calorie. It is striated blue and white: ice and air. This ice is ancient, made of compressed snows that were laid down, high on the mountains above, centuries ago. Its entombed bubbles contain a few molecules of air that were last breathed, perhaps, by Magellan or Vespucci.

Seven minke whales are feeding at the base of Endurance Glacier. They rise among the ice floes, breathing explosively once or twice, and then silently sound. After each whale dives, it leaves a footprint on the sea surface. They are feeding on the krill beneath the broken ice. A front of fog is advancing across the bay like an opaque veil. It envelops the whales, but through the fog I can still hear their breaths. The icebergs loom briefly in the gray-purple fog and disappear, changing from cerulean to gray to invisible. The bergs are grounded, but the fog is moving. Briefly they reappear, like phantoms in a realm that is no more substantial than water and water vapor.

Now the fog bank has stopped between the ship and the shore, halted by the katabatic zephyrs from Endurance Glacier. In the late afternoon, some of us take a Zodiac, a motorized, inflatable boat made of rubbered cloth, to Lookout Point. In a few minutes we sail from gray dusk to clear twilight. Three Weddell seals are basking on a rotten ice floe, brown with feces. They ignore us. A leopard seal swims to the Zodiac, peering at us with his huge, somewhat reptilian head, which seems to be three-quarters mouth. The sea is crossed with flocculent reddish brown streamers of guano expelled by the penguins as they streak toward shore. Yesterday this stuff was krill; now it is plankton food and will soon become diatoms and krill once more in the frenzied ocean metabolism of summer. Now we can see the ocean floor, this biome that is one with the deepest hidden areas of the world. But the shallows are nearly lifeless, a pavement of rock that bears pale wounds inflicted by icebergs. Not much is able to survive in waters less than ten meters deep on the Antarctic continental shelf, which is incessantly bashed and scraped by ice. Only beneath the shelter of overhanging ledges and rocks are there tufts of green algae and a few Antarctic limpets.

Lookout Point is a small mountain that is joined to Elephant Island by a narrow beach of pebbles. A few kilometers away, behind the fog bank at the western end of Endurance Glacier, is Cape Vigilant, where in 1918 Ernest Shackleton and his crew of twenty-seven bivouacked after their ship, the *Endurance,* was crushed in the ice of the Weddell Sea. Leaving the others behind, Shackleton and five others left Cape Vigilant for South Georgia Island, 1,290 kilometers northeast across the ice-strewn sea, in a 6.7-meter-long open boat. It was a daring move that eventually led to the rescue of all his men. Since they departed, the only visitors to Elephant Island have been the occasional scientific party and, recently, tourists on expedition cruise ships.

Several dozen elephant seals, wallowing on the beach in pools of their own feces and penguin guano, watch indifferently as we land. They have watery, bloodshot eyes and runny noses. A few raise their heads and gutturally protest our arrival; their voices are like massive, prolonged belches. But the early springtime days of reproduction, the maniacal defense of territory and harem, are long over. The pups are already made this year, and there is nothing to do but molt, rest, and eat plenty of squid in preparation for next season's squabbles and trysts. The beach is littered with white penguin skins, turned inside out and knobby with the follicles that seat the feathers. They were the victims of leopard seals, which grab a penguin at sea and, by repeatedly slapping the carcass on the water, shuck the body from its skin. There is also the corpse of a cattle egret, a tropical bird that eats grasshoppers and other insects kicked up by the hooves of ungulates. It obviously had no business venturing to Antarctica.[8] Rocky stretches of beach are covered with empty limpet shells regurgitated by the Dominican gulls, which swallow them whole. Above the reach of the tide lies a freeze-dried elephant seal. It may be many decades old. It appears to be part of the rock itself, and indeed it is as hard as rock. The skin has worn off its ribs and spine, which are starkly white. From each of its vertebrae emerges the delicate white webbing of its nervous system, which seems to resist decomposition and which spreads across the basalt as if enervating it.

Behind the beach is a cliff face of black basalt. Orange lichens encrust certain facets of the cliff, but others, of different geometry and orientation, are bare. Pintado petrels nest on the rock ledges. They sit on their eggs like contented hens, waiting to be relieved by their mates who are feeding at sea. Wilson's storm petrels also nest among the rocks. In flight they keep to the lee of the peninsula, fluttering like dark moths over black rock and then tumbling into the wind at the edge of the cliff.

* * *

After a stormy night passage we arrive at Admiralty Bay. The *Barão de Teffé* is anchored in the pastel shallows in front of the station. The tedious transfer of equipment and personnel takes all day. During the night, when the wind whipped across the rolling deck of the ship and strummed her cables and lanyards, a snow petrel collided with a communications antenna. It was one of the young of the season, just starting its life of sea and air, of following the ice edge in seasonal migration with the sun. It must have suffered unilateral brain damage, for its left foot is withered and it favors its right wing. A compassionate crew member, who fed it ham and black beans and kept it locked in the salt-water shower with the cold water running, handed me the crumpled little bird. Holding it is like touching animate, warm ice.

I keep this bird in the food freezer. Obligatorily a denizen of the cold, it can't survive indoor temperatures. How alien to my physiology is this little creature, forever bound to the frozen pole, as I am forever imperiled by it. How contrary we are as life forms. My first task is to purge it of the ham and beans. Four times a day I push smooth yellow amphipods — crustaceans the size of peas — down its gullet with my right index finger, much as its parents must have done with their blunt bills. And then I give it seawater, dripped from a finger. Afterward, for several hours, its tube-nose copiously excretes salty fluid, the snotty sign of good health. In the morning and afternoon I exercise it on the wind-washed hills behind the station. At best, it glides a meter on the soft cushion of air just above the ground and then crashes. There are rain puddles in the swales. On cold mornings they are frozen, and this bird of ice, not understanding ice and thinking it liquid, skitters and slides in confusion across the clear, hard puddles and crashes again. It has realized that I mean it no harm when I frighten it into the air, and now it refuses to budge. Sometimes I hold its body and force its wings to flap in the cold air. But it has gotten no stronger. How can I take care of this white alien? Should I release it? Grounded, it won't survive long, but nor does it have a future living with humans. Today a brown skua, also born this season, alights a few meters away and patiently watches the petrel flail. It has the calm patience of a scavenger. After the skua leaves, I drop the petrel into the sea. It floats on the calm water that it can never leave, tucks its feathers, and paddles away.

2
Memories of Gondwana

The woods decay, the woods decay and fall,
The vapors weep their burthen to the ground.
— ALFRED, LORD TENNYSON

A WIND IS TAPPING sleet on the double-paned window by
my bunk. It is an erratic wind, rapidly changing direction
and thrust, as if trying to find a chink in the green sleeping
module. On this first dawn in Antarctica, around two-thirty in
the morning, I contemplate the thin membrane of metal and glass
that separates me from the immense crystal desert outside. The
station is like a ship surrounded by a hostile sea. I, a relatively
hairless primate whose ancestors evolved in tropical Africa, have
no business being here. The curtain that separates my bunk from
the others provides the only privacy I will have during my stay in
Antarctica; the challenge I will face this summer will not be to
combat the elements but to peacefully cooperate with my neighbors. I jump down from the bunk and walk through the meteorology module, through the dining lounge to the kitchen. The
station is asleep. During the long summer day it chooses to conform to the alien metabolism of the north. In the kitchen the
cafezinho, as always, is strong and sweet. Nearby is the hot foyer,
where our outdoor clothing is dried and stored. The foyer is
Antarctica's threshold, like the airlock on a submarine: one al-

ways stops there after entering and before exiting. I suit up in the
foyer, relishing the dry warmth of my down jacket — my surrogate fur — and step out into Antarctica.

The first thing I notice is the special clarity of light. The sleet,
which has blown in a dark eddy down the bay, has washed the
beach, the rocks, the snow, even the air. The air has truly become
invisible, a phenomenon that is unknown in the temperate and
tropical realms of dust storms, humidity, pollen, and pollution. It
is as if I have suddenly acquired the vision of an eagle. Flagstaff
Mountain, six hundred meters above the station, seems almost
touchable. I can perceive every detail of fault, dip, and strike
etched in the mountain slopes two kilometers across the bay at
Ullman Spur. I can trace every striated snow shelf and crevasse of
the Ajax Icefall and Stenhouse, Goetel, and Dobrowolski glaciers.
Angrily puncturing Dobrowolski Glacier, parting its flow and
creating ice falls on its flanks, is the Ternyck Nunatak. Some of
the crevasses are luminescent turquoise, as if they were leaking a
hidden energy; they are, in fact, only reflecting this fresh morning
light. Their cerulean color is caused by the filtration of the longer
wave lengths — those toward the red end of the spectrum — on
the light's long journey through the reflecting facets and fractures
of the interior of glacier. Only the more penetrating blues escape.

Flotillas of icebergs sail across the bay, pushed by the same
north wind that blew the sleet away. They have dropped from the
glacial faces at the head of the bay and will bounce along like
sluggish pinballs, crowding one shore or the other, sometimes
becoming grounded, until they either melt, trailing plumes of
sediment in the water, or work their way out of the mouth of the
bay to be borne off by the currents of Bransfield Strait. They don't
melt very rapidly. The bay is poised on the margin between liquid
and solid. Its temperature is 2° C and neither the darkest nights
of winter nor the hottest days of summer budge it very far from
that point. The bay is chalky turquoise, as if strangely pigmented.
It doesn't seem to be sea water but some sort of liquid mineral,
which, in fact, it is. Its color is derived from the suspended glacial

flour, produced and released by the icebergs, which, like the glacial ice, differentially absorbs the long-wave radiation, leaving only the blue.

The station is on the edge of Keller Peninsula, a tongue of rock two kilometers long and one kilometer wide. The interior of the peninsula is a spine of volcanos one hundred and fifty million years old. In their youth, these peaks were magma cores, the hot oozings of heat generated by the collision and subduction of tectonic plates. Today they are humpbacked and hoared with ice. The crumbling mountain tops occasionally cast shards onto their slopes and onto the beach below: volcanic bombs that look like petrified cow dung, ice-shattered fieldstones like loaves of sliced bread, copper-imbued bluestones, brick-red ironstones. Some of the mountains are flanked by seasonal snow fields, tinged with pink algae at this time of the year, which on hot summer afternoons dribble ephemeral streams onto the beach below. Late in the afternoon, when the sun dips behind the mountains and the chill once again returns to the mountain slopes, the streams congeal and become silent.

The station is built on a long, rambling beach of pebbles cast down from the mountain above. From here we can safely walk two kilometers south toward Plaza Point and another three kilometers up the western flank of the peninsula without encountering glaciers, crevasses, whiteouts, disorientation, and the other pervasive hazards of Antarctica. With a little planning, we can even hike over the tumbled scree slopes of the hills behind the station into narrow valleys with close horizons of rock and ice. We can take the Zodiacs into the relatively safe waters of the bay to tend our nets and to collect specimens on other shores. But there is always danger. Sprain an ankle, break a leg, or be knocked unconscious by a falling rock, and you can die in the cold before being rescued. Afflicted with engine failure and nudged by a stiff northern wind, a Zodiac could drift with the ice out of the mouth of the bay into Bransfield Strait. Minor mishaps can quickly grow into catastrophes; the four crosses on the hilltop behind the station attest to this. So there are rules. Before hiking beyond sight of the station or taking a Zodiac, we have to enter in a journal our names, time of departure, destination, and expected time of return. We must never travel alone. Even in the wilderness there is no privacy. We must take walkie-talkies and check in at designated times, and if we don't maintain contact, a search party will be sent after us. Search parties have been formed needlessly when communication has lapsed because of a frozen battery or when an intervening mountain has prevented radio transmissions. But on several occasions they have saved lives.

*   *   *

Our task today is to change the tape of the data recorder at the automatic weather station and to replace the burned-out light bulbs on the crucifix at the summit of Flagstaff Mountain. This involves a 600-meter climb. As a biologist, I have no particular business as mountain climber or electrician, let alone as keeper of the faith, but I volunteer because I want to witness the bay from above. The three climbers are led by Helena, who is the *alpinista,* or mountaineer, for our team. During the winter she is a kindergarten teacher in a poverty-stricken *favela* of São Paulo. I would never guess that this quiet woman spends the rest of the year scaling peaks in the Andes, the Alps, and now Antarctica. In the billiard room she teaches me the ropes — the use of crampons, the ice axe, fast knots, and slip knots — with the same gentle patience that she must use with her students.

The climb begins in the low hills behind the station, past the four graves, past the freshwater lake that provides our drinking water, and onto the wind-scoured slopes of the center of the peninsula. A few gyres of fast wind, remnants of this morning's squall, unpredictably tumble down the slopes. Tatters of ice fog cling to the peak of Flagstaff Mountain, but the surrounding glaciers are sun-dashed. Although the valley floors are barren, the ridges are festooned with foliose lichens, some of which curl like pale green leaves ten centimeters tall. Why do they prefer these exposed edges where the wind batters them the most? Some of the boulders, lichen-encrusted, look as if paint had been spilled on them, or as if they were spattered with bird droppings. A few have been shattered by alternating frost and thaw, and the crustose lichens that grow on top of them have shattered as well.

The escarpment is dynamic on this warm, rainy, sunny, windy morning, as if the mountain were living. It thawed yesterday, then froze, then snowed, and it's thawing again today, creating all manner of ice and snow. The earth is both solid and liquid. It harbors last winter's snows, frozen sand, and rubble and is bruised by oozing mud slides in which you can sink up to your knees. I understand this morning why Eskimos have dozens of words for snow and as many more for ice. As temporary ice dams succumb to the late morning heat, pulses of streaming water, bearing slush and grit, rattle down the smooth ice slopes. Occasionally black boulders are also released by the melt. They bound down the slope, spinning in the air, skipping into the swales. One of these fast rocks could easily kill a person.

We walk halfway around the mountain and then back again before deciding how to ascend to the peak. We cross an eroded glacier, sideways like crabs, laboriously cutting steps in its flank with our ice axes until, an hour later, we reach the base of the summit, where the scree and mud are piled precariously high. The final slope is just at the angle of repose, beyond which any movement — any footfall or handfall — dislodges a shower of debris. Here the mountain face is laminated with rock and mud — I can't tell earth from dirty ice. The terrain is so steep that I use the ice axe as a metal claw to pull myself higher, a reach at a time. At one point in the middle of the tilted plain, my calf starts to cramp, and I use the embedded ice ax as a foothold. It is all that prevents me from falling onto the rocks below.

Helena, with a grace born of confidence and experience, continues to the summit and drops a line to me. She wraps it around her waist and braces her boots against an old volcanic core. Safety is above, and I pull myself higher. The summit has the characteristic sharp spires of ice and thaw. From here I can peer across the glaciated plateau of King George Island, the generator of clouds and the tumbling katabatic winds that push the icebergs around the bay. The restive bay is as multihued as an artist's palette and, like the rocks, its character is defined by a shifting margin of freeze and thaw. The water fronting the beach is turbid with muddy runoff, the slow dissolution of the mountain on which I am standing. The glaciers are also dissolving, filling the bays with brash ice. They groan and shift and occasionally calve a berg: Antarctic thunder.

We switch the tapes on the weather recorder and walk over to the six-meter-high crucifix. The old bulbs have been shattered by wind-blown particles of ice and sand. A nest of wires feeds the bulbs, and a long cable tumbles down the cliff face to the station generator. After replacing the bulbs, we eat lunch: peanut butter and jelly sandwiches with *guaraná,* a caffeine-rich Brazilian beverage made from the fruit of an Amazonian vine. It tastes like honey and water.

In a temperate climate, this mountain would be the aerie of an eagle or a condor, or would have a meadow of alpine flowers. But this wind, which could easily waft a large predatory bird, is barren. Even at this modest height in Antarctica there is nothing living save for a few hunched lichens and perhaps a few microbes. The biosphere, the mantle of life that so exuberantly cloaks Earth, has shrunk to a thin film over this spot, and in a morning's climb I have left it. From this promontory King George Island looks as prevital as the moon: the gray faces of glaciers, gray sky, a few black sinews of rock. Far below on the thin beach is the station, itself like a small green lichen in an incomprehensibly vast wilderness.

*   *   *

The descent is easier: we toboggan on our backsides down the snow face and onto the glacier. Since we are traveling at the speed of the falling rocks, they are no danger to us, but we must hustle to run clear of them once we stop. Safely below the zone of falling rocks, we decide to return via the northern flank of Flagstaff Mountain, down a slope of volcanic scree mixed with older sedimentary rocks. It is here that I notice a particularly colorful rock banded in yellow and violet. Shards of it are scattered in a little heap, and I reconstruct them like pieces of a puzzle. Slowly a pattern of concentric rings becomes discernible and I realize that I have found a piece of petrified wood. The rings are alternately purple and thin (indicating the months of winter dormancy) and pale and wide (summer growth). It is clear that even the mountains are mutable here. The barren flanks of Flagstaff Mountain must have been forested long ago.

It is a disturbing thought, but I am hardly the first to have this realization. Since humans began to voyage to the Antarctic Peninsula a century and a half ago, they have been puzzled by locally abundant, and seemingly displaced, fossils embedded in the rocks: petrified tree trunks, leaves, even the occasional fossil flower. The first fossil recorded from Antarctica was found by James Eights, a physician from Albany, New York, who from 1829 to 1831 served as naturalist aboard the sealing vessels *Penguin, Seraph,* and *Annawan.*[1] Nathaniel Palmer, at age thirty-one already an accomplished Antarctic explorer, was the leader of the expedition. While walking on a slope like this one, probably somewhere on King George Island, Eights found "a fragment of carbonized wood, imbedded in . . . conglomerate. It was in a vertical position, about two and a half feet in length, and four inches in diameter; its colour is black, exhibiting a fine ligneous structure; the concentric circles so distinctly visible on its superior end."[2] In Eights's day, fossils were still great mysteries. Many believed them to be remnants of the Great Flood described in the Bible, and therefore irrefutably younger than 4004 B.C.

Eighty years after Eights's discovery, fossil trees were found on the other side of Antarctica, in the Olive Block region of the Transantarctic Mountains. There, on February 8, 1912, while returning from the South Pole, Robert Falcon Scott, Edward

Wilson, "Birdie" Bowers, Titus Oates, and Taff Evans paused at the top of Beardmore Glacier, 1,200–1,800 meters above sea level, to collect sixteen kilograms of geological specimens, including fossil wood. Since early November they had been trekking across the Ross Ice Shelf and the glaciers of the Antarctic Plateau. "A lot could be written on the delight of setting foot on rock after 14 weeks of snow and ice and nearly 7 out of sight of aught else," wrote Scott in his journal.

> It is like going ashore after a sea voyage. . . . We decided to steer for a moraine under Mt. Buckley and, pulling with crampons, we crossed some very irregular steep slopes with big crevasses and slid down towards the rocks. The moraine was obviously so interesting that when we had advanced some miles and got out of the wind, I decided to camp and spend the rest of the day geologising. . . . We found ourselves under perpendicular cliffs of Beacon sandstone, weathering rapidly and carrying veritable coal seams. From the last Wilson, with his sharp eyes, has picked several plant impressions, the last a piece of coal with beautifully traced leaves in layers, and some excellently preserved impressions of thick stems, showing cellular structure.[3]

Wilson and Scott were determined to bring the fossils to the attention of the world. This heroic scientific decision may have contributed to the failure of the expedition. During the long days of sledge-hauling down the frozen cascades of Beardmore Glacier and over the undulating Ross Ice Shelf, Evans and Oates succumbed to frostbite and exhaustion, and the chances that the three survivors would safely return to the base camp on Ross Island daily grew more remote. Yet Scott, Wilson, and Bowers doggedly continued to carry the fossils. On March 29 all three men froze to death while confined to their tent by a blizzard, only eighteen kilometers from a depot of fuel and food. The following November the three corpses, covered by their collapsed tent, were found by a rescue party. Close by was the sledge, loaded with fossils.

Decades passed before it was learned how ancient these fossils were. The Olive Block fossils are from trees that grew on the fringe of an inland sea 2.5 to 3.2 million years ago. Yet they are among the youngest Antarctic fossils. Eights's fossils, as well as the one I found on Flagstaff Mountain, were from the Miocene, 16 million years ago. They have survived 16 million cycles of winter and summer, yet still bear the mark of life. Both the King George Island and the Olive Block fossils are of southern beech trees *(Nothofagus)*, a genus found today in Patagonia, New Zealand, Tasmania, Australia, New Caledonia, and New Guinea.

The fossils implied wondrous things: that Antarctica has not always been ice-locked and that it was once attached to the continents in the north. Clearly, Antarctica is not prevital; it is a formerly living continent that has been stripped of its epaulets of life and exiled to the bottom of the world. Life's diversity has been distilled to a few simple forms, not by heat but by cold. This insight, which slowly evolved for over a century, revolutionized geology. We now know that before 200 million years ago, Antarctica was part of a much larger continent, posthumously named Gondwana, that also included what is now Australia, India, Africa, Madagascar, and South America. Gondwana lay in the middle latitudes of the Southern Hemisphere and had a benign, perhaps even subtropical, climate.

The oldest Antarctic fossils, tracings of marine invertebrates dating from the Cambrian 500 million years ago, are nearly as ancient as the vanished continent itself. By the Permian, 225 to 280 million years ago, entire terrestrial plant communities had been preserved in the fossil record. The most detailed Permian fossils, from Mount Augusta in the Transantarctic Mountains, are the silicified wood of arborescent gymnosperms, some of which grew as tall as twenty-two meters. The fossils also tell us of their environment and of life's struggle. The annual growth rings of one of these tree species, *Araucarioxylon,* are highly variable, ranging from negligible to eight millimeters wide, implying that the trees were subject to fluctuations of climate during their lives, perhaps heavy springtime cloud cover, drought, or autumn frosts. The wood also shows "false rings," ephemeral decreases in growth rate, perhaps because of infestation by insects. Some of the fossils are scarified and healed, implying damage by fire. Others show clearly preserved fungal hyphae, which created necrotic pockets deep inside the heartwood.

The next glimpse of the Antarctic past is provided by fossils dating from the Triassic, 25 million years later. They were also discovered in the Transantarctic Mountains, on the eastern edge of Beardmore Glacier, preserved in the sediments of a long-vanished stream bed. A few shards of tooth and bone, including a fragment of the jaw of a meter-and-a-half-long labyrinthodont amphibian, which resembled a salamander, are all that have been recovered. One hundred sixty kilometers away, in the Coalsack Bluff of the Alexandra Range, preserved in deposits of similar age, are the remains of *Lystrosaurus,* an aquatic reptile a half meter to a meter and a half long. The nostrils and eyes of *Lystrosaurus* were in turrets on the top of its head, enabling it to breath while submerged. The genus is thought to be a characteristic fossil of Gondwana; other *Lystrosaurus* fossils from the same epoch have been found in India and southern Africa.

Abundant fossils, dating from the Jurassic, 190 to 135 million years old, have been found at the tip of the Antarctic Peninsula on the slopes of Mount Flora (a misnomer, since during the Jurassic there were no flowering plants). Here, preserved in golden iron pyrites, are the impressions of subtropical vegetation — ferns and their primitive allies — as well as freshwater fish scales, snails, and the impressions of aquatic beetles. The Mount Flora fossils, numbering several dozen taxa, were first collected during the summer of 1901–2 by members of the Swedish South Polar Expedition. Since then, tons of specimens have been dug from the deposit, but few, if any, new species have been discovered. This impoverished biota probably existed on the edge of a freshwater lake subjected to periodic inundation, a condition that fostered fossilization but diminished species richness.

The south polar region of Earth was just as dark in the winter during the Jurassic as it is now, but there was no continent, just ocean. That was soon (at least as measured in geological time) to change. Starting about 150 million years ago, tectonic forces began to tear Gondwana asunder. The breakup was slow, only a few centimeters per year, but inexorable. Africa separated first, setting off on a path that nudged it against Eurasia, where plants and animals from the north invaded repeatedly. (Madagascar, which separated from Africa about 100 million years ago, was spared these invasions and as a result still supports a unique and relict Gondwana biota.) A few million years later India broke away from Gondwana, drifted northward across the equator, and crashed into central Asia; the crumpled edge of the collision became the Himalayas. South America, Australia, and Antarctica were still connected at the beginning of the Cretaceous, 120 million years ago. What remained of the fractured continent had a hospitable climate; its southern shore was highly seasonal, and its northern edge was subtropical. Gondwana remained stationary in these amiable middle latitudes for nearly 50 million years. The seas that surrounded the continent harbored species that have since disappeared: pleiosaurs (aquatic dinosaurs, fifteen meters long) and ammonites (shelled relatives of squids) as well as sharks and lobsters (extant in other parts of the world today, but absent from Antarctica). But these were also seas of genesis, where the ancestor of the penguins lost its ability to fly in the air and entered the ocean.

By the mid-Cretaceous, about 100 million years ago, flowering plants, including the characteristic southern beeches, began to appear on Gondwana. Dinosaurs were at their peak, and mammals, particularly marsupials, were becoming ever more diverse and ecologically important. When the dinosaurs abruptly went extinct about 66 million years ago, mammals became the dominant vertebrate animals. In 1981, Michael Woodburne and William Daily, paleontologists at the University of California at Riverside, found a 40-million-year-old fossil jaw of an opossum,

Polydolops, on Seymour Island, east of the Antarctic Peninsula in the Weddell Sea.[4] This small, furtive animal probably subsisted on berries and insects. It was perhaps the most important paleontological discovery ever made in Antarctica. For more than a century zoologists have been mystified by the disjunct distribution of the marsupials, which are found in Australia (kangaroos, wallabies, koalas, wombats, and the extinct predatory marsupial "wolf," *Thylacine,* to name a few) and the Americas (a diversity of opossums, the extinct doglike borhyaenids, and even a saber-toothed predatory marsupial). The oldest known marsupial fossils, which are from North America, are 100 million years old. How did the marsupials get from North America to Australia? Until the Seymour Island discovery, theories ranged from crossing the Bering Sea land bridge (unlikely, since there are no marsupials in Asia) to rafting across the Pacific on logs or floating mats of vegetation (extremely improbable for a large mammal). But now it is clear: marsupials *walked* from North America to Australia via South America and Antarctica.

✳   ✳   ✳

Much of the folding of Antarctica took place during this time. The ice and the deep sea conceal the details of the process. The peninsula resulted from the collision of tectonic plates along the same subduction zone (where one plate slides beneath another) that crinkled the western coast of South America and formed the Andes. As in the Andes, the collision of plates generated enormous volcanism, and much of the topography of the South Shetlands, the South Orkneys, and the South Sandwich Islands is the result of that forge.

Antarctic volcanism continues to this day. The early explorers found it incongruous — and sublime — that fire should accompany ice. Upon seeing Mount Erebus, an active volcano on Ross Island, Joseph Dalton Hooker wrote in 1840,

This was a sight so surpassing every thing that can be imagined, and so heightened by the consciousness that we have penetrated, under the guidance of our commander, into regions far beyond what was ever dreamed practicable, that it really caused a feeling of awe to steal over us, at the consideration of our own comparative insignificance and helplessness, and at the same time an indescribable feeling of the greatness of the Creator in the works of his hand . . . and then to see the dark cloud of smoke, tinged with flame, rising from the volcano in a perfect unbroken column; one side jet-black, the other giving back the colours of the sun, sometimes turning off at a right angle by some current of wind, and stretching many miles to the leeward![5]

Hooker was the surgeon and naturalist on the British National Expedition to Antarctica, which consisted of the three-masted Royal Navy warships *Erebus* and *Terror,* from 1839 to 1843. The expedition was under the command of James Ross, who had joined the Royal Navy at age eleven, was already a veteran of four expeditions to the Arctic Sea, and had in 1831 been first mate on a voyage that successfully located the north magnetic pole. Ross was perhaps the most experienced polar explorer of his time.[6] Now determined to find the south magnetic pole, Ross embarked on one of the epic explorations of all time. He visited Kerguelen Island, Tasmania, Campbell Island, and, in two consecutive summers, dodging 1,600 kilometers of broken pack ice, reached the southern and western shores of the sea now named after him. He was the first to describe the great ice shelf also named in his honor. "Never did I meet with such devotees to science," wrote Hooker of Ross and his crew. "Captain Ross's little hammock swings close to his darling pendulum [used to measure gravitational anomalies], a large hole in the thin partition allowing him to view it at any moment." The expedition, declared Hooker, "cannot fail to prove of inestimable value to science in its various departments, and to maintain, for the British Navy, that pre-eminent rank it has so long held among the nations: 'terrible in war,' and during times of peace."[7]

Things were not as sublime in 1893 when C. A. Larsen, a Norwegian explorer and whaler, landed on Zavadovskiy Island in the South Sandwich group and was overcome by volcanic fumes seeping from cracks in the ground. He was debilitated for several months afterward. In the South Shetlands, Deception Island and Bridgeman Island have active, but sporadic, volcanism. When, in 1822, James Weddell, a Scottish sealer and explorer, visited Bridgeman on the brig *Jane,* there was so much "smoke issuing through fissures of the rock and apparently with much force" that his ship could sail no closer than 180 meters from the island and could not land.[8] In 1880, when another party of British sealers landed on Bridgeman Island, they killed a fur seal whose fur had been singed by hot lava. But the volcanism was fickle; when J. B. Charcot sailed past the island in 1909, there was no activity.

✳   .✳   ✳

King George Island is a confusing patchwork of igneous and sedimentary rocks of various ages, most laid down in the past 50 million years. In Admiralty Bay, the eastern end of Dufayel Island has an abundance of Eocene fossil plants, 40 to 50 million years old. A kilometer and a half away, at the mouth of Escurra Inlet, on the other side of a submarine fault, are younger fossil plants from the Lower to Middle Oligocene, 30 to 38 million years old; and eight kilometers east-northeast, between Lion's Rump and Low Head, are volcanic rocks bearing layers of marine conglomerate with fossil mollusks and foraminiferans dating from the Pliocene, less than 5 million years old. But the most evocative fossils on King George Island may be the eighteen-centimeter-long footprints of a large bird. Although they are 46 million years old, they seem to stride across the mudstone as if made yesterday. These are probably the tracks of the so-called terror bird, a flightless, fast-running relative of the cranes and rails that stood three and a half meters tall. Its clawed feet enabled it to disembowel mammals, including the two-meter-long glyptodonts, which were armored like armadillos. Terror birds probably originated in South America, where they may have hastened the extinction of large marsupial carnivores. At the time of the closure of the Darien Gap, 2.5 million years ago, they were the largest carnivores in South America.

On the eastern shore of Admiralty Bay, at Cape Hennequin, there is a fossil forest: 16-million-year-old fossil impressions of plants embedded in ripple-marked and aqueous tuffs. These include leaves of several species of *Nothofagus* (tinged brownish as if they had just fallen in the autumn wind), of the gymnosperm genus *Araucaria*, of ferns, and of several unnamed flowering trees that drifted into a fast-flowing stream and were rapidly smothered by sediments. My fossil, which so eloquently depicts seasonality in its stone rings, is from this epoch, when Admiralty Bay had a climate and flora nearly identical to those of Patagonia today. Imagine: flowing rivers and warm, halcyon summers. Did the leaves of this tree flutter in a wind that blew off a temperate sea? Did birds forage in its boughs by day; did marsupials snuffle in its leaf litter by night? Was its pollen dispersed by the wind, or was it pollinated by long-vanished insects?

✳   ✳   ✳

The virtue of this planet, which distinguishes it from most, if not all, of the rest of the cold, indifferent universe, is life. Earth would be better named Vita. As a biologist, I celebrate life — such obvious employment, given where I live. Most of my research has been in the rain forest of the Amazon, which is the most diverse and complex biome to have evolved during the 3.5-billion-year history of life on Earth. There, steeped in life's exuberance, I try to decipher the reasons for that diversity — how so many species have come into being and have managed to coexist over the millennia. Terrestrial Antarctica, of course, is the antithesis of the Amazon, which may be why it so appeals to me. It is like the silence between movements of a symphony. But it is also a key to the diversity of the tropics. The Antarctic is the weather factory, the generator of winds and ocean currents, which may have, through an unexpected series of events, driven speciation in the distant tropics.

The process began with the final fracturing of Gondwana about 50 million years ago, when South America separated from Antarctica, and 10 million years later, when Australia sheared away, stranding Antarctica in the cold and gloom of the South Pole. Surrounded by the Southern Ocean, which was widening at a rate of a few centimeters per year, Antarctica cooled at a rate of about 1° C every million years. Eventually snow began to accumulate faster than it melted, and during the mid-Pliocene, about 4 million years ago, the present-day ice sheet began to form. Antarctica became a continent arrested in adolescence by the waxing sea and cold. The life of the interior of the continent was slowly rubbed out by glaciers, and the surviving life forms crowded to the maritime margins or to offshore islands. The Southern Ocean became colder and storm-tossed. Unimpeded by land masses, the West Wind Drift slowly strengthened and began to generate the cyclonic storms that are so characteristic of the region in our time. Antarctic surface water began to flow north, forming the Convergence, and the Antarctic bottom water carpeted the seafloor. The effects of these changes were felt all over Earth.

Recently conditions have become much more severe. During the past 1.8 million years there have been at least four, and perhaps as many as ten, ice ages. The last one (known as the Wisconsin Ice Age in the New World) was at its peak only 19,000 years ago. In North America the ice cap extended as far south as Iowa and Illinois; in Europe it covered the British Isles. In Antarctica the glaciers projected hundreds of kilometers out to sea, obliterating the shoreline, where life was crowded, and crushing most of the continental shelf. The South Shetland Islands, from Deception to King George, were fused under a single cap of ice. So much water was tied up in the glaciers that sea levels dropped by about ninety meters all over the planet. Fortunately the ice ages were short-lived, at least from the perspective of geological time. Most lasted only 30,000 or 40,000 years, and in the periods of thaw between the glaciations, the glaciers melted and sea levels rose higher than they are today, flooding many of the continental shorelines. The evidence of most of these episodes of thaw has been obliterated by erosion, but on King George Island the melts left raised beaches, up to twenty meters above the present-day shoreline, bearing the shells of marine mollusks and other invertebrates. The Brazilian Station is built on such a shelf.

During the peaks of glaciation, near-freezing Antarctic surface and bottom water engorged the cold ocean currents that ply north along the western shores of Australia, South America, and Africa, changing patterns of atmospheric circulation. The result may have been the desiccation of all three continents. Western Australia, already a desert, became colder and drier. The pampas and the coastal deserts of the southern third of South America crept north, and the Amazon forest retreated into a dozen or so small refuges surrounded by arid savanna. In Africa the Kalahari Desert edged as far north as the Congo River Valley and, like the Amazon forest, the Guineo-Congolean forest withered into a few isolated pockets. The tropical forest refugia became virtual islands, with little or no genetic exchange. Species that once had been widespread were broken into several isolated populations. In each population the particular genetic traits of the founding individuals and the particular selective pressures of each forest island fostered the divergence of daughter species from their common ancestor. When the glaciers melted, conditions of humidity returned to the Amazon and Congo river valleys. The forest once again became a continuous belt, and the ranges of the sibling species overlapped. If little evolutionary divergence had taken place between taxa in the interim, they may still have been sufficiently related to hybridize. If they were genetically incompatible and could not hybridize, yet made similar demands on their environment — for food, a place to grow, or light, for example — then one sibling species, over time, may have competitively excluded the other. But if they were sufficiently different so that competition was slight or nil, then they may have managed to coexist, and what had once been a single species now became several. Moreover, this process was exponential, because each of the sibling species split into as many new species as there were refugia during the next cycle of glaciation. The episodic dissection and merging of the two great zones of continuous tropical forest, choreographed by climatic events in the new Antarctic continent, may be an explanation for their extraordinary species richness.

In a strikingly similar process, Pleistocene refuges may also explain the surprising richness of the sessile benthic fauna of the Antarctic continental shelf and the affinity of that fauna with forms of the deep sea. As the glaciers crushed the Antarctic continental shelf, shallow-water marine forms had to retreat beyond the reach of the glaciers — or perish. In most areas their only escape was into submarine canyons that extended into the abyssal plain to the north. Like the forest refugia 6,000 kilometers to the north, these canyons became virtual islands, and their benthic faunas became isolated. Species that once had been circumpolar were broken into many small and isolated populations. When the glaciers retreated, the Antarctic continental shelf rebounded (but because of its long compression by ice it is still the deepest shelf of all the continents), and the sibling species recolonized the empty shallows. As in the tropics, their diversity multiplied exponentially.

While the Pleistocene refuge theory may explain, at least partially, the evolution of the enormous richness of species of the tropical forests and the Antarctic continental shelf, it cannot explain the *maintenance* of diversity. One way that species richness could be maintained, at least in theory, is by prolonged competitive exclusion, meaning that an organism's chances of dying because of competition from individuals of a closely related species are less than those of dying because of some outside force. In tropical forests an example of competition-independent mortality occurs when a falling tree destroys other trees growing beneath it, as well as those woven to its canopy by vines; in the Amazon forest, up to 60 percent of the trees die because they are clobbered by a neighbor. Such a high rate of competition-independent mortality has a profound effect on the forest ecosystem: it prolongs the time necessary for one species to drive a closely related one to extinction, because in the course of its lifetime an individual is likely to succumb to a treefall before the selective pressures of competition can operate on it. On the Antarctic continental shelf, competition-independent mortality is caused by rocks, ranging in size from pebbles to massive boulders, dropped by passing glaciers that scour debris from the mainland. The rocks create a mosaic very similar to the patterns of treefall in a tropical forest and, as in a tropical forest, their falling may foster species richness.

In both the equatorial tropical forests and the Antarctic continental shelf, adversity — glacial catastrophes, killer trees, and rocks dropping from above — spawned, and maintained, a bouquet of new species. However, in terrestrial Antarctica, during the ice ages most life was obliterated, although some simple life forms — a few species of bacteria, algae, perhaps a few mites — may have taken refuge on ice-free nunataks. The plants and animals we observe today are mostly late arrivals that have colonized the margins of the continent since the glaciers retreated, bobbing ashore on flotsam, carried by the wind, or hitching rides on the feet and feathers of birds.

And in a continent of latecomers, I am the last. My species brings its own shelter, energy, and johnny-come-lately comforts; the crucifix, its bulbs replaced, tonight burns strong in the short night. Naked, clothed in stolen fur, we visit during a brief respite from the ice. And unless we as a species somehow addle the tick and cycle of climate and glacier, all this will once more be scraped clean, the gouge that is Admiralty Bay will deepen, and the life that hauls itself out on the shore or coats the seafloor will once again retreat beyond the edge of cold or into the submarine canyons.

3
Life in a Footprint

I could be bounded in a nut-shell,
and count myself a king of infinite space . . .

—SHAKESPEARE

A BOATLOAD OF TOURISTS visited the station today. Most are geriatric — the working young can't afford junkets to Antarctica — and many are widows. What curiosity, pent up over a lifetime, has led these retired schoolteachers and housewives to this most remote place on Earth near the end of their days? One woman told me that she wanted to glimpse Antarctica as a last act before she died. She wanted to see the uncluttered face of the world. Did she find solace here? She shuffled through the station, exchanging trifles with the personnel and taking snapshots, then wandered past British Point to Cousteau's whale skeleton and took more snaps. Striving to get just the perfect angle, she trampled the moss bed around the skeleton. Her footprints were like oozing wounds. Then, climbing the slope behind the station, she headed for the four graves. With one misstep, the woman sent a cascade of rock and crumpled lichen into the valley below. She may not have known it, but she left her indelible mark; the continent will long remember her brief passage.

*   *   *

On the leeward edge of Keller Peninsula, facing west, in the shelter of frost-splintered boulders, are patches of green mosses, bundle grasses, and Antarctic pinks. The patches are no bigger than cloud shadows and just as variable. All of these plants, but especially the pinks, hunch from the elements, adopting the hemispheric cushion shape that exposes the least surface area to the desiccating wind, and it is moist in the interstices of the stems and leaves. The plants grow on a thin soil of crushed rock and guano and, on this warm afternoon, are watered by melting snow. Broken limpets dropped by Dominican gulls have decomposed into a rich organic soil. I lie on my belly, scrutinizing one of these patches, only a few meters wide, observing every iota of life: every mite clambering over the shiny green filaments of moss, every blade of grass, the curve of a fruticose lichen.[1]

The Antarctic pink and Antarctic bundle grass are the only two flowering plants in the Antarctic.[2] By contrast, there are over a hundred species of flowering plants in the comparable north polar latitudes and, in life's greatest eloquence of diversity, the Amazon Basin, there are fifty thousand species. The patch I am examining is as complex as Antarctic terrestrial ecosystems get. Consider this: one leaf of an Amazonian palm — one leaf out of the trillions in that 3,000-kilometer-wide swatch of tropical forest — may have living on top of it more species of mosses, fungi, lichens, protozoans, mites, and insects than are found on the entire continent of Antarctica.

Both pinks and bundle grasses are common in Patagonia, and the seeds of their ancestors probably arrived in Antarctica as hitchhikers on the feathers of birds. Tufts of bundle grass are a favorite nesting material of south polar skuas and Dominican gulls. But the obstacles to the plants' survival in Antarctica are nearly insurmountable: drought, cold, cryoturbation (the polygonal cracking of the soil by alternating freeze and thaw), an absence of insect pollinators, and long months of twilight and darkness. And in only a few clement areas of Antarctica, where soils are moistened by glacial meltwater and fertilized by birds, where the terrain faces the sunny north and is close to the modulating sea, do the two species grow together. Keller Peninsula is one of those areas.[3]

Recently scoured by glaciers, Admiralty Bay has thin and scruffy patches of mosses, like blots on an artist's palette. But parts of the South Orkneys, South Georgia, and several sub-antarctic islands are muffled in thick deposits of peat, some of which are nearly four meters deep. Only the top layer of the peat is living moss; the rest is the accumulated organic material of generations past, which decomposes slowly because of the cold and anaerobic conditions of the bog. The bottom of the peat beds on South Georgia have been radiocarbon-dated at 9,700 years old; on Macquarie Island they are 10,275 years old, and on Kerguelen Island 11,010 years old. Obviously these plant communities started growing as soon as the glaciers retreated at the end of the last ice age.

But most of Antarctica seems embalmed in ice and only briefly, if ever, thaws. Even here in the Banana Belt, the mosses and lichens of Keller Peninsula must withstand being covered by snow for nine months of the year. The snow algae, single-celled plants that colonize the surface of the snow and derive their nutrients from wind-blown dust, tinge the snow red and, by increasing the absorption of solar radiation, hasten the springtime thaw. To get an early start in the spring, the bundle grasses begin photosynthesis while still covered by snow, and their photosynthetic metabolism is fully stoked at light levels that would starve a temperate plant.

It may seem paradoxical that most of Antarctica, where ice and snow are so pervasive, is a desert. But for the plants and animals that live here, liquid water is the rarest of resources. The yellow crustose lichen *Verrucaria serpuloides* solves the problem by growing on submerged rocks in the sea. The bay is a far more benign place than the air: it is wet, has a stable temperature, and is washed by nutrients from the penguin rookery at Point Thomas. Some terrestrial Antarctic lichens grow on top of moist moss beds, robbing their hosts not only of water but of light. (In response, the mosses manufacture antibiotic substances that prevent epiphytic lichens from becoming established.) But most terrestrial lichens tough it out, and when desiccated they simply shut off their metabolisms and enter dormancy. When moisture returns, they are able to revive in hours, removing water vapor directly from the atmosphere and increasing their weight many times on a humid day.

Fruticose lichens are the dominant vegetation on the rocky slopes of the South Shetland Islands, tinting the landscape a soft gray-green. Their blades curl in the dry wind and expand in the wet, and the texture of the hillsides seems to change with the weather from day to day. These wisps of life seem as insubstantial as the blowing snow, but recent studies have showed them to be some of the most enduring organisms on Earth. The science of measuring the weight gain of lichens over short periods (usually no more than several months or years) and using these rates to estimate their age, is known as lichenometry. The extrapolation is fraught with assumptions: that lichens don't swell when they take up water, that local environmental conditions — such as the orientation of their substrate relative to the sun, its warm sides and cold sides, or the proximity of a nutrient-rich penguin rookery — have no effect on growth rate. But even if one assumes a large margin of error, the results are startling. For example, a typical adult *Usnea antarctica,* a fruticose lichen that is common on Keller Peninsula and that attains a maximum length of 25 centimeters, grows at a rate of about 0.45 millimeters per year. Simple math tells us that the larger individuals could easily be more than 500 years old. The largest recorded specimen of *U. antarctica* was 67 centimeters long, suggesting an age of about 1,500 years. Yet among lichens, these are the young whippersnappers. Individuals of the Antarctic crustose lichen *Rhizocarpon geographicum,* which grows only 10 millimeters in diameter per century, may be 4,000 years old. But most lichens live short, fast lives (at least on a lichenometric scale) of decades or less. *Usnea fasciata,* a close relative and neighbor of *U. antarctica,* grows three times faster than its congener, and the encrusting *Buellia latemarginata,* which is abundant on coastal rocks, increases its weight by up to 33 percent per year. Among individuals of *U. antarctica* that are less than five years old, mortality is nearly 100 percent per year, although among those few that survive to become centenarians it is nearly zero. Most of what I see are improbable survivors, and from the scale of my lifetime they seem immortal, as I am to the springtails and midges. Imagine: these lichens, which if I am not careful I plod over with indifference, are as old as the Egyptian pyramids, older than the language that I speak and write.

✳   ✳   ✳

On this liquid December afternoon the sun is strong and lingering, and the glaciers behind the beach are melting. The mosses, pinks, and grasses are like sponges. Each leaf and filament wicks the ephemeral liquid and supports a film of bacteria, yeasts, algae, and protozoans. A bright red orobatid mite, less than a millimeter long, is swilling on this scum. Innumerable springtails, primitive wingless arthropods that can flip themselves into the air like fleas off a cat's back, also inhabit the lichens and mosses. Not much bigger than grains of sand, springtails can walk on the surface tension of water, and they are so numerous — tens of thousands per square meter — that the melt pools seem to have a living skin. The springtails are in the shadow of a rock and are nearly moribund with cold. But I can rouse them to activity by merely exhaling on them; my metabolic fires are a fine substitute for the sun. Nearby, in a slanting shaft of sunlight, are two *Belgica,* flightless midges that are the most southerly free-living insects on Earth. The tips of their abdomens are welded together in amplexus, the male squeezing his seed into his mate and leading her, in short centrifugal steps, to the transient warmth. Still joined, they walk onto my finger. At two millimeters long, they are giants by comparison to the mites and springtails. Giant, indeed: *Belgica,* this mote of a creature, is the largest terrestrial animal on the Antarctic continent.[4] It is the invisible animals — the protozoans that live in films of water — that are the most abundant and have the greatest biomass. Theirs is the economy of being one-celled; through binary fission they are able to reproduce in minutes and quickly take advantage of the summer thaw; at the onset of winter they encyst just as quickly. *Belgica* feeds on this film of organisms. It is at the top of a food chain that is only two links long: no predators eat *Belgica* (although no doubt some parasites demolish it from within). Its principal enemy is invisible: the incessant and drying wind, which would blow a winged midge to its death in the nearby sea.[5]

The urgent union of the two *Belgica* may last for several days; in a land of scarcity, it's adaptive to hang on to a good thing. For these little animals, it is not only favorable habitats that are in short supply; often mates are as well, and sex, which through genetic recombination produces a bouquet of genetic expression, is as essential for long-term survival in a changing environment as food and water.[6] Coition completed, the female *Belgica* will survive just long enough to lay her gelatinous egg mass. Before the summer's warmth ends, the eggs will hatch into brown maggots that swarm over the moss beds and the filth of the penguin rookeries in densities as high as several thousand per square meter. Many of the maggots will die in the winter, when the air temperature (although modulated by the nearby sea) drops to as low as $-15°$ C.

Although water, like light, is a precious and limiting resource for Antarctic animals, it becomes deadly when it turns to ice. Water has the peculiar property of expanding when it freezes, and cells (which are mostly water) tend to rupture when ice crystals form inside them. Antarctic terrestrial invertebrates employ three general strategies to adapt to the cold. The tardigrades (eight-legged arthropods that live in aquatic or moist environments) simply dehydrate; ice can't form where there is no water. The larval midges submit to the cold and freeze, a cryogeny that would kill most other animals. They survive because ice crystals form in the spaces *between* cells, not *inside* them, so that the delicate intracellular machinery is not disrupted. The springtails and mites employ a more sophisticated strategy to survive the winter temperatures. They produce biological antifreezes, such as sugars and polyhydric alcohols, which permit their body fluids to remain liquid at subfreezing temperatures, a process known as supercooling. But a supercooled fluid is a hair-trigger medium, in which ice can instantly precipitate around a nucleating particle such as a fleck of mineral dust. Therefore springtails and mites prepare for winter by fasting and completely evacuating their guts of any debris.

The best strategy for avoiding desiccation and cold is to hide inside of, or snuggle close to, a warm-blooded animal. Many Antarctic terrestrial invertebrates are hitchhikers or parasites on the skin and feathers of birds or in the nasal passages, genitals, and alimentary tracts of seals. Every species of Antarctic bird has its own species of biting louse, and some bird species are infested by as many as three. In spite of their name, the biting lice do not suck blood but scrounge a living by eating feathers and dead skin. Fleas do suck blood, and six species infest Antarctic seabirds. Five of the fleas are cosmopolitan species; only one, a parasite of the silver-gray petrel, is endemic to the continent. Each species of true Antarctic seal — leopard, crabeater, Ross, and Weddell — has its own species of sucking louse. These relationships indicate a long evolutionary association between parasite and host, a slow pas de deux spanning millennia. For a flea or a louse, the radiant, nutritious alien being to which it clings is the universe. The parasite ever accommodates its ways to the evolutionary meanders of its host; the host mounts defenses that overwhelm it or force it to adapt once more. These noxious freeloaders have followed the retreating southern continent to the bottom of the world, migrating with the seasons to and from the edge of the cold. But it is no small feat to hang on to the feathers of a petrel while it plunges into the freezing sea, to hide in the wet anal and genital folds of a diving Weddell seal or, like the elephant seal louse, to excavate and huddle within an air-filled sheath in scaly epidermis while being squeezed by 30 atmospheres of pressure at a depth of 300 meters.

＊ ＊ ＊

The first explorers in Antarctica had their senses tuned to a planetary scale. Their accomplishments were heroic: the crossing of oceans and the discovery of new lands. Hunters of seals and whales, they took little notice of the scruffy plants and minute fauna between their boots. James Weddell, the Scottish sealer and explorer, noted the presence of bundle grass while visiting Laurie Island in the South Orkneys in 1823. But it was not until the summer of 1829–30 that the species was properly collected and described, by James Eights, who published the description of "an occasional plant of a small species of avena" in 1833. Soon after, one of Eights's specimens was sent to William Hooker, the celebrated English botanist and father of Joseph. The elder Hooker, astonished by the tenacity of the little plant, lamented, "What, it may be asked, can be expected in the way of Botany, in those dreary regions of the extreme south, where the rigour of the climate and the striking diminution of vegetation . . . appear to offer an effectual barrier to the very existence of plants?" Regardless, Joseph enthusiastically accompanied James Ross to the Antarctic to document its impoverished flora. "Our young naturalist," wrote William of his son, "declares that mental occupation afforded him the sole relief from the anxieties and ennui incident on the voyage."[7] Ross encouraged his officers, and especially young Hooker, to collect and describe the plants, marine fauna, and geology of the new lands, and his collaboration with Hooker was one of the most creative partnerships in all of discovery.

Although the expedition reached the southern edge of the Ross Sea, Hooker's most interesting observations were not of Antarctica itself but of the subantarctic islands. During the austral summer of 1840, the expedition spent two months in Christmas Harbor on Kerguelen Island, a snowcapped speck in the middle of the southern Indian Ocean, while Ross put the expedition's flock of sheep ashore to graze on the abundant tussock grass (beginning a slow defoliation of the island that continues today). Hooker was at liberty to roam the treeless island. "My rambles were generally solitary," he wrote, "through the wildest country I ever beheld. The hills were always covered with frozen snow, and many of my best *Lichens* and *Mosses* were obtained by hammering at the icy tufts, or sitting on them till they thawed." Hooker recognized the aerodynamic economy of the cushion plants. "All these plants," he wrote, "whose stems rise a little above ground, are flexible, and bow beneath the blast, while the chief part are of lilliputian growth, and form such dense and interwoven masses, that the very soil must flee away in dust, ere they could quit their position."[8]

Hooker was particularly intrigued by the Kerguelen cabbage, a half-meter-tall relative of cultivated cabbage. Although vitamin C would not be discovered for over a century, by Ross's time the value of fresh and pickled vegetables (especially sauerkraut) to combat scurvy had been well established. The stores of the *Erebus* and the *Terror* were provided with an abundant supply of cranberries, pickles, and mustard, but mariners were ever searching for new, local antiscorbutics. Kerguelen cabbage seemed ideal. Wrote Hooker, "The root tastes like *Horse-radish,* the seeds like those of *Cress;* but the leaves are the grand fresh provision, and were so extremely relished by the sailors, that during the whole of our sojourn in that barren land, they were always boiled with the ship's company's beef, pork, or pea-soup. They taste to me very like very stale cabbage, with a most disagreeable essential oil. . . . This oil gives this cabbage a curious anti-heartburn property. Altogether, I consider this cabbage a most valuable antiscorbutic, which few persons do not like, or cannot bring themselves to eat."[9]

Two years later the expedition visited the Falkland Islands. The Falklands, like most subantarctic islands, have vast moors of tussock grass, an exuberant species that grows a meter tall and creates massive clumps of intertwined roots.[10] Unlike the diminu-

nental shelf, was scraped nearly clean of life repeatedly during the Pleistocene glaciations, and the last glacial maximum, 18,000 years ago, may not have been the most severe. Could any terrestrial plants and animals have survived the glaciations? Certainly microbes could have survived. Viable bacteria locked in frozen sediments at least 10,000 years old have been recovered from cores of soil and ice 76 to 427 meters deep on Ross Island. The frozen soils near the hut constructed in 1917 by Robert Falcon Scott, also on Ross Island, have yielded viable *Escherichia coli* that may have lived in the intestines of the expedition members. And the soft-bodied mites, which today are abundant on ice-locked nunataks (indeed, one species has been discovered on a mountaintop only 5° from the South Pole) could have taken refuge on the tops of nunataks that were ice-free during the ice ages. But the greatest refugia may have been seasonal ponds on top of the ice itself, examples of which are extant today on the Ross Ice Shelf, an undulating ice plain that is frozen for most of the year. During the hot, bright day of summer, ephemeral melt ponds appear on the ice surface and are rapidly colonized by dense mats of algae, bacteria, cyanobacteria (some bright green and orange), tardigrades, and even nematode worms. The ice shelf moves constantly, and the mats are always changing shape. Cracks open and drain the ponds, which then form somewhere else. Inevitably the dried mats ablate and blow away, spreading plant and animal propagules downwind for hundreds of kilometers, as far away as Victoria Land. It is the ideal life strategy for the Antarctic: living on top of the ice and dispersing in the wind.

But the bacteria, mites, and pond dwellers would have been exceptions; probably little else survived the Pleistocene in Antarctica. Most of what we see today are recent colonists that have arrived in Antarctica since the end of the ice age 10,000 years ago. Wind — the dominant motif of Antarctica — is the obvious vector. The Antarctic atmosphere is a broth of microscopic propagules — bacteria, flecks of algae, the spores of mosses and lichens — that are as invisible as the air that carried them from across the sea, and beyond. There are even occasional records of much larger wind-borne animals: a spider that was trapped at Marble Head (77° S), and a mite that was filtered from the atmosphere 900 meters above sea level halfway between New Zealand and Antarctica. However, since most winds close to Antarctica flow north, wind dispersal must have been an infrequent event, probably on the southward eddies of cyclonic storms. And once airborne, the colonists' troubles were just beginning. They had to survive the desiccation and extreme cold of high altitude, and the chances were negligible that they would drop on favorable terrain at the right time of the year. More likely, the ancestral fauna and flora were carried south stuck to the feet and feathers of migrating birds; several springtail eggs and thousands of moss spores can fit under a scale on an albatross's foot. A few species, such as the moss *Polytrichium alpestre* and the fruticose lichen *Cladonia rangiferina,* are found in both the Arctic and the Antarctic, and could have crossed the torrid zones only as hitchhikers on the bodies of bipolar birds, such as the Arctic tern.

✳ ✳ ✳

Located in the Banana Belt of Antarctica, the Antarctic Peninsula and the South Shetland Islands are relatively hospitable places. Conditions for life are far more severe on the continent itself. There, even during the summer, life has pervasive, ominous marginality. One is never quite comfortable, and the sun, although constant, never warms. I once spent an afternoon in early March on the shores of Terra Nova Bay (74°21′ south), on the western edge of the Ross Sea under the hot brow of Mount Melbourne, which, like Mount Erebus, is an active volcano. The gray sandstones of Terra Nova Bay were swept clean of lichens and mosses by the dry katabatic winds that slithered to the shore from the glaciated interior. Terra Nova Bay is at the northern end of the valleys, known as "dry valleys," of the Admiralty Range. They are quite unlike any other places on Earth.

The dry valleys are Antarctic oases, but they are not conventional oases of moisture in a desert. Quite the contrary. They are oases of bare rock and soil in an ice-smothered crystal desert. During the Pleistocene the valleys were repeatedly gouged by glaciers, but today the fringing mountains obstruct the flow from the glacial plateau, and dry winds sublimate ice and snow before it melts. The 5,000 square kilometers of dry valleys in southern Victoria Land are the largest ice-free areas on the continent. The Wright Valley has Antarctica's only river, the Onyx, which seasonally drains the glacier at its head. A few animals have straggled in here, but the trip is one-way; in the lower Taylor Valley, 160 kilometers from the sea and next to a perpetually frozen pond named Mummy Lake, are the freeze-dried remains of seals. They lie as they died: no skuas or giant petrels have scavenged their innards, although radiocarbon dating has revealed one of the seals to be over a thousand years old. Like every place in Antarctica, the dry valleys are dusted with a variety of microorganisms, none of which seem able to establish a foothold. But appearance is deceiving. In the dry valleys, where the grainy sandstones have been cleaved by the heave of frost and thaw, are the tracings of lichens and cyanobacteria that have insinuated themselves into the crystalline interstices of the rock itself. The endolithic organisms are environmental refugees, able to survive only in translucent rocks — quartz, granite, granodiorite, and certain marbles — that permit the passage of sunlight, and since these organisms do not actively dissolve minerals, they can colonize only rocks that are porous. Most endolithic organisms enter only the shallow surface fissures and cracks, but the endolithic lichens invade channels and cavities inside the rock itself, to a depth of several millimeters. The lichens are layered black, white, and green (a stratification that also occurs in surface-dwelling lichens). All three layers are permeated with the hyphae of filamentous fungi, but the photosynthetic green algae cells — the precious engines that enable survival — are concentrated in the deepest, most sheltered zone. Only when the rock erodes, often foliating in the cutting wind, do the pieces of fungus and alga disperse. Most of the propagules fall onto unsuitable areas and fail, but a fortunate few may lodge in a sheltered notch or a fortuitous fissure of rock and begin to invade.

The biota of the dry valleys is the ultimate in ecological reductionism. These are the simplest of Earth's ecosystems: the sunfixing algae and the fungal and bacterial decomposers, all within a few millimeters of each other. There are no consumers or predators. By comparison, the moss and bundle grass beds of Admiralty Bay have a labyrinthine complexity.

The dry valleys also have luxuriant beds of moss, but they are hidden beneath the frozen lakes. Freshwater lakes, nourished by melting snow and ice, are common in coastal Antarctica. Some are permanently frozen; others are seasonally liquid. The lakes surrounding Admiralty Bay, like many ponds in the Antarctic Peninsula and its nearby islands, support microscopic protozoans, tardigrades, and rotifers, and the macroscopic copepods that feed on them. But the frozen lakes of the dry valleys are among the most alien places on Earth. They are up to forty-six meters deep and are perennially covered by as much as three meters of ice. During the winter the air temperature above the lakes drops to $-51°$ C, but beneath the ice the lake is always liquid, and the bottoms of some lakes, even in the darkest days of winter, may be as warm as $22°$ C. During the spring and autumn most of the sunlight is absorbed or reflected by the covering snow. Yet vertical crystals of ice have a certain light-gathering efficiency, acting as fiber optic cables that channel the summer sunlight into the liquid in sufficient amounts to support exuberant populations of bacteria, green algae, and aquatic mosses. Like the endolithic lichens, they could not survive exposure to the atmosphere, and also like the lichens, have taken shelter inside a solid, in this case ice instead of rock. As the frozen lake surface is sheared and sublimated by the wind, new ice replaces it from the water layer beneath. This conveyor traps strands and flecks of living material, which slowly work their way upward until they reach the surface, where they are desiccated and scattered by the wind.

In 1976 the *Viking* landers found no evidence of life on Mars.

Instead they found an arid planet, with a thin atmosphere that permitted vast fluxes of ultraviolet light and with seasonally waxing polar caps of frozen carbon dioxide. But there is no reason to declare that Mars has been abiotic throughout its history. Both Mars and Earth formed about 4.6 billion years ago. Earth, and perhaps Mars as well, was warm and wet during its first 500 million years and may have had huge equatorial bodies of water. The first inklings of life — bacteria and cyanobacteria — appeared on Earth at this time, and their descendants have continued irresistibly ever since, populating the entire face of the planet, transforming the atmosphere with oxygen, even reworking its surface geology. But early in its history Mars, for reasons that remain inscrutable, succumbed to the cold of its distant orbit from the sun, and its atmosphere dissipated into space. Any life that the planet had nurtured was destroyed. There are abiding questions. If life existed on Mars, how long did it linger as the twilit planet cooled? Did it take refuge under stones, under frozen lakes? If it did, it must have been very similar to present-day Antarctic bacteria and algae, which are just hanging onto the edge of an inhospitable continent. Earth has abundant fossils of her earliest life: stromatolites (columnar depositions of rock laid down by cyanobacteria and algae) 3.5 billion years old from Western Australia and Southern Africa. Perhaps Mars does as well. Can these living Antarctic rocks teach us what to seek when finally we (or our robots) return to the Red Planet?

4

Penguins and Hormones

To live is like to love — all reason is against it,
and all healthy instinct for it.

—SAMUEL BUTLER

I N A STIFF NORTHWEST wind you can smell the chinstrap
penguin rookery at Bailey Head, on the eastern side of Deception Island, from sixty kilometers out to sea. It is a fecal
barnyard stench, the excreta, in a bountiful summer, of perhaps
180,000 nesting penguins. As one approaches to within a few
kilometers of shore, the sea broils with thousands of chinstraps
weaving through the swells like plump flying fish. Some muster
sufficient momentum to actually fling themselves between the
wave crests and briefly become birds of airborne flight. Others
streak like dolphins in the blue sea or ride the bow waves of the
infrequent passing ships. This is not a summer frolic but an urgent
and dangerous gauntlet: the sea is hungry with penguin-eating
seals and orcas, also converging on Bailey Head. The ocean turns
with eddies of waterlogged feathers and the bobbing, empty skins
of penguins turned inside out by leopard seals.

At the black sand beach the penguins belly-surf on the last
breakers and flop ashore, waddle, and flop again as the next wave
bashes them from behind. Here at last it is safe, and they rest on
their bellies or stand stolidly amid a flotsam of vesicular pumice

and corniced bergy bits — the fractal shapes of volcanism and summer melt. On this afternoon there are other animals washed ashore: a few medusae, their bells as transparent and as blue as the sea, and gelatinous ribbons of salps (pelagic tunicates) that slowly freeze-dry in the wind.

The penguin rookery is in a volcanic crater separated from the beach by a lip of ash sixty meters high. A seasonal stream has cut through the ash. On this warm afternoon in January, when the temperature has edged to just above freezing, it is a torrent. The penguins waddle through the ravine on leveed banks of ash, and I join them. Those returning from the sea, their feathers shiny and clean, mix with those leaving the rookery, their feathers guano-spattered and ruffled from the hassles of chick-rearing. The living waddle indifferently past the slowly decomposing bodies of the dead: scraps of chicks torn apart by skuas, a few adults that have struggled ashore after sustaining injuries at sea. A juvenile elephant seal has also died in the ravine. Two giant petrels stick their heads deep into its body cavity and enthusiastically gulp little putrid morsels of tissue. They defend themselves by ejecting their gorge, consisting of their most recent meal and a fetid stomach oil, at their enemies, and they lurch at me as I pass, bills gaping.

Entering the rookery, I am surrounded by a cacophony of bickerings and gabblings and a billow of stench. I have the sensation of entering a packed stadium. What a confusion it is: penguins dusting the hills to the horizon, ninety thousand guano-splatted nests. But to a penguin this is not a visual panorama, nor can they decipher patterns in the olfactory waves. It is acoustical. The babel I perceive is eloquence to a penguin. Each returning penguin has memorized the particular timbre and pitch of its mate's voice, and each baby knows its parents' voices. An enormous amount of information is being exchanged here: I'm hungry; I'm horny; I belong to you; I've survived the hungry sea; here I am.

The crater floor is a giant sewer, with pools of guano — some bearing coiled tapeworms fifteen centimeters long — broken eggs, upchucked krill, and bright green patches of *Praseola,* a short-lived alga that thrives in the burning nitrogen of excreta. An elephant seal has undulated ashore and indifferently flopped on the penguin nests, crushing the eggs and scattering the chicks. The penguins that nest here have little success. The most experienced penguins climb to the sun- and wind-swept flanks of the crater, where the soil is clean and dry, and some trek as far as three kilometers to its far lip, pulling themselves with beak and claw up steep slopes. But it is a long haul, weaving through their neighbors' nesting territories and enduring their sharp little pecks.

*   *   *

Ice mimics rock on Deception Island, a volcanic crater that rises 70 meters above the sea 145 kilometers south southeast of King George Island. Repeated eruptions have transformed and buckled the landscape, creating volcanic cones and pushing sediments with embedded marine mollusk shells 60 meters above sea level. Summer rains have mixed the layers of ash and dirty snow. The most recent eruptions, between 1967 and 1970, smothered the last clumps of bundle grass, but already the skuas and Dominican gulls have brought in new propagules, and several new species of mosses that have never before been recorded in the Antarctic now grow near the fumaroles of those eruptions.

From afar the island appears to be steep-walled and inhospitable, with few beaches and no safe anchorage. But the southeastern lip of the crater, known as Neptune's Bellows, has collapsed, and the sea floods into the heart of the island. During spring tides the current rips through the opening at a speed of one and a half knots. Inside is a vast and safe anchorage with steaming fringes where geothermal activity has melted the snow and, in a few cul-de-sacs, turned the sea scalding. Here thousands of brittle stars, fleeing the poaching sea, writhe onto the beach and dry in the frozen wind, and krill wash ashore fully cooked and bright red, providing a lazy feast for the Dominican gulls. Just to the north of Neptune's Bellows is another hole in the wall, known as the

Window. The pintado petrels nest on the cliff face of the Window, finding it easy to launch into the funneling wind.

During November 1820, the American sealer Nathaniel Palmer discovered the secret heart of Deception Island. The chinstraps must have coursed under the bow of his sloop, the *Hero,* just as they do under my Zodiac, 165 years later. The caldera may be the best anchorage in all of Antarctica, and in the century and a half since Palmer's discovery it has been used repeatedly by explorers, whalers, scientists, even aviators. Nine years after Palmer's discovery, Captain Henry Foster, in command of HM Sloop *Chanticleer,* conducted an extensive exploration of Deception Island. A record of the journey was written by W. H. B. Webster, the ship's surgeon. Not only was the expedition high adventure, it was the epitome of science for its time: using pendulums placed in various parts of the Northern and Southern hemispheres, the explorers measured variations in Earth's gravitational field. The *Chanticleer* spent fifty-four days anchored in the flooded caldera (since named after Foster), in a bay that still bears the name of Pendulum Cove. But the island wore heavily on Webster's spirit. "A more dreary or more cheerless scene cannot be imagined," he wrote, "than that which Deception Island of Shetland presented. . . . Here all is joyless and comfortless, huge masses of cinders and ashes lie strewed about, which imagination converts to the refuse of Vulcan's forge . . . hills of black dust and ashes topped with snow, and enormous icebergs buried beneath immense loads of volcanic matter."[1]

Webster observed that the seeping gases and freezing temperatures of Deception Island induced a peculiar sterilizing effect and arrested decay. He noted a total absence of colds and other respiratory diseases among the crew. "It was found that putrefaction does not readily take place [in this] climate," wrote Webster, "for on opening a grave [of an unknown sealer] . . . the body was found entire, and free from any unpleasant odour, although we had reason for believing that it had lain there some years."[2]

The crew of the *Chanticleer* made regular forays over the

southeastern lip of the caldera to Bailey Head to capture penguins for the cooking pot. "But though some thousands [of penguins] might be numbered among the slaughtered, salted and eaten, by our people," wrote Webster, "I question whether we made any apparent diminution of their numbers." Webster griped about the taste of his quarry.

> Although an edible food, I cannot say much for the flesh of the penguins that we obtained at Deception Island in this particular. The method of cooking them, by which we found they were most palatable, was that of frying them in slices with a little port; but when they were salted, and preserved for any length of time, they certainly form the most disgusting food, worse even than dog-meat of the most repulsive kind."[3]

The melancholy Webster must have walked on this beach where I walk today. He described Bailey Head as "a dreary and gloomy waste covered with penguins innumerable, the noise of which was not unlike the distant honking of folded flocks of sheep. . . . Instead of the variety of birds of elegant plumage which adorn other happier regions, hosts of penguins here strut about with stupid mein, harmless and happy in their dreary abode as they are unsuspicious of harm from man."[4]

∗   ∗   ∗

Since the discovery of Antarctica, our species has abused its nonchalant penguin neighbors. Having evolved in a realm without terrestrial predators, penguins have no instinctive fear of humans, and submit innocently to club and gun. Their meat and eggs long have been a staple food for humans and sled dogs. And in their sacrifice, penguins have on several occasions saved the lives of explorers. During the winter of 1903, the Norwegian explorer and whaler Carl Anton Larsen and his shipwrecked crew subsisted for an entire winter on the inhabitants of a nearby Adélie penguin rookery on Paulet Island. Four years before, it was Frederick Cook's idea to extract his ship, the *Belgica*, from the pack ice by lubricating her hull with the fat of penguin hides, and his innovation saved the expedition. As recently as the 1980s, personnel at Argentina's Esperanza Station fed penguins to their dogs.

But it is the industrial hunting of penguins that historically has been most destructive. During the nineteenth century uncounted millions of penguins, mostly rockhoppers and kings from the subantarctic islands, were rendered into oil to feed the insatiable needs of the Industrial Revolution. On Macquarie Island, shortly after the arrival of the first sealers in the 1820s, hundreds of thousands of king and royal penguins, mostly infants of the season, were herded onto inclined ramps like compliant sheep and forced to leap into cauldrons of boiling oil. The industry stopped only after the penguin populations were reduced to levels beyond which it was not economical to hunt them. Thirty years later in the Falkland Islands, 400,000 penguins were bludgeoned to death, skinned, and cooked in a single year. Although the Falklands have abundant deposits of combustible peat, the try-pots were fired with penguin skins, which bear a great deal of subcutaneous fat. The yield from this enterprise was 50,700 gallons of oil, or only about a pint per penguin, but the aftershocks of the industry are still noticeable in the diminished populations of penguins today.

✳   ✳   ✳

There are eighteen species of penguins, all from the Southern Hemisphere and all from Gondwana. Only two — the Adélie and the emperor — are truly Antarctic, spending their entire lives on the continent or close to its shores. The emperor, which breeds in the Antarctic winter on top of fast ice, is a wholly oceanic species — indeed, the only species of bird that need never alight on land in the course of its life. The other species of penguins that nest in the Antarctic — the chinstrap, gentoo, king, macaroni, and royal penguins — have ranges that include the subantarctic islands.[5] These seven species comprise 70 percent of the avian biomass in Antarctica.

King George Island is as cosmopolitan a place for penguins as any on Earth. Chinstrap, Adélie, and gentoo penguins all breed at Point Thomas and nearby Sphinx Hill, at the western entrance of Admiralty Bay. The three species are close relatives, all members of the family of "brush-tailed" penguins. In comparison to Bailey Head, this rookery is modest: only 18,000 Adélies, 2,800 gentoos, and 300 chinstraps. At first it would seem that they all make their living the same way: crowding onto the same low hills and beaches, vying for the same finite number of nesting pebbles and feeding on the same populations of krill in the bay. Careful long-term study at Point Thomas has revealed important differences in the life styles of the three species, which have carved distinct temporal and spatial niches.[6] It is only because Point Thomas is neither extremely Antarctic nor clemently subantarctic that they manage to coexist there. The Adélie penguin is adapted to the hostile conditions of the Antarctic continent and is near the northern edge of its range. The gentoo, a subantarctic species that is common as far north as the Falkland Islands, is at the southern edge of its range. Only the chinstrap, a species of the maritime Antarctic, is truly at home. The social structure of each species is woven into instinct and is adapted to the exigencies of place and time. The chinstraps have a strong perennial fidelity to both their nesting sites and their mates. This loyalty is probably due to the inclement and variable conditions of their breeding sites along the Scotia Arc: a known nesting site and a tested mate are advantages in an unpredictable world, but only if the breeding season is long enough — and benign enough — to find another partner should last season's consort not return in the spring. By contrast, the southerly Adélie penguins retain an attachment to their nest sites but not necessarily to their mates of the previous season; time is too short for the luxury of fidelity. Like the chinstraps, the gentoos maintain a strong pair bond over the seasons, but having evolved in a relatively warm sea with a surfeit of nesting beaches, they waste little energy squabbling over real estate.

The reproductive sequences of the three species at Point Thomas are separated by intervals of several weeks. The first ashore, in late September, are the Adélies. The gentoos arrive in mid-October and the chinstraps three or four weeks later. Courtship, egg laying, and the nurturing and fledging of the chicks are therefore staggered; when the gentoos are at the peak of hatching in middle December, the Adélie chicks are two weeks old and the chinstraps are still incubating. Competition among the penguin species for krill — especially during the demanding period when the chicks are being nurtured — is therefore diminished. The three species also reduce competition by having different foraging ranges. Gentoos are homebodies that stay inside Admiralty Bay or close to its mouth in Bransfield Strait. Chinstraps range as far as 34 kilometers into the Strait; Adélies up to 50 kilometers. Their vertical ranges are different as well: gentoos dive to 150 meters, which explains why they can catch sufficient krill in Admiralty Bay; chinstraps and Adélies make much shallower dives, to a maximum of about 100 meters.

*   *   *

The origins of penguins have long been debated by ornithologists. At the turn of the century, developmental biologists mistook flightlessness for primitiveness and concluded that penguins were a missing link between dinosaurs and birds. The proof of this hypothesis, it was thought, lay in the anatomy not of the adult but of the embryo, since ontogeny was believed to recapitulate the evolutionary stages of the penguin ancestors. For this reason penguin eggs were sought by ornithologists; the egg of the emperor penguin, which breeds during the Antarctic winter, was the most elusive prize of all. The adventure that led to the recovery of the first emperor eggs, by Edward Wilson, "Birdie" Bowers, and Apsley Cherry-Garrard, is one of the great sagas of exploration. On June 27, 1911, the three departed from Robert Falcon Scott's hut at Cape Evans on a trek to the emperor penguin rookery at Cape Crozier, 105 kilometers east, on the opposite side of Ross Island. They traveled on skis, hauling their equipment in sleds. It was the dread middle of the austral winter, and on some days temperatures dropped as low as − 54° C. The wind was just as remorseless. On one occasion it ripped the team's only tent from its moorings and carried it 300 meters away; fortunately the explorers were bivouacking in a nearby igloo. Dysentery, frozen sweats, hypothermia, and incessant uncertainty plagued the explorers. Toward the end, they were so delirious from exhaustion and exposure that they repeatedly had to nudge each other awake. "This I know," wrote Cherry-Garrard, "we on this journey were already beginning to think of death as a friend." On July 20 the explorers reached the penguin rookery on the fast ice in front of the looming sixteen-meter-high face of the Ross Ice Shelf.

> We saw the Emperors standing all together huddled under the Barrier cliff some hundreds of yards away. . . . After indescribable effort and hardship we were witnessing a marvel of the natural world, and we were the first and only men who had ever done so; we had within our grasp material which might prove of the utmost importance to science; we were turning theories into facts with every observation we made, and we had but a moment to give.[7]

Thirty-six days later the team staggered into the hut at Cape Evans with three emperor eggs, which were later dissected at the University of Edinburgh. But the embryos did not provide the proof that the explorers were seeking; penguins were in fact very modern birds. The following summer Wilson and Bowers both froze to death on the Ross Ice Shelf, while returning from the Pole with Scott.[8]

We know now that penguins are very much creatures of Gondwana. Their origin is documented by abundant fossils, including the remains of a giant penguin that stood nearly two meters tall and weighed thirty-two kilograms, from Seymour Island. Hardly primitive, penguins are recent descendants of flighted birds, probably coast-dwelling relatives of petrels that took to the sea about 40 million years ago, after Antarctica had split from the northern continents and become isolated in the Southern Ocean. As Antarctica slowly drifted south and became ice-locked, its resident penguins adapted to the increasingly hostile conditions. The penguins invaded a niche — the open and highly productive sea to a depth of several hundred meters — that was out of reach of all other birds. The evolution from flight to swimming involved a striking reversal of several adaptations that characterize flighted birds. Penguins lost their hollow, air-filled bones and evolved solid, ballasted ones. The wing bones, which in flighted birds are so adroitly and flexibly articulated, are in the aquatic penguin fused into a stiff flipper, a rudder-wing that is rigid enough to push through the heavy medium of water. In contrast to all other birds, their wings are designed to push down, not up; when they stop beating their wings, penguins bob upward and crash on the surface. Indeed, to say that penguins are "flightless" is misleading, for penguins do fly on their modified wings, only their medium is water, not air.

It is during the warm summer day that we observe penguins. They are land animals then, nesting in clamorous colonies, scrapping for two-dimensional patches of real estate and small pebbles. But the major portion of the penguin's life is spent in the dark winter sea. Watching nesting penguins on this sunny afternoon at Bailey Head, I can barely imagine their three-dimensional seafaring life: the incessant energy of the waves, their urgent dives to peck krill from the water, followed by explosive draughts of air and long rests on drifting ice floes. Theirs is a twilight realm under the sunless winter sea and ice, but a realm of clarity and dimension, since under the muffling ice, the Southern Ocean precipitates its sediments and is air-clear. They must experience constant anxiety about leopard seals and orcas. Do they remember the warm summer idylls, the sun, their anxiety about skuas?

The enduring quality of penguins is instinct. The monotonous clamor of the penguin courtship displays is hard-wired and unvarying. They are also beasts of chemical impulse. In an instant, during the deepest, darkest nights of the austral winter, the hypothalamuses of 35 million chinstrap penguins each secrete a few molecules of sex hormone. And at that moment it becomes inevitable that when the light returns to Antarctica, the inanimate scree valleys and snow fields of continent and island will resound with the gabble of displays, and the hills will be streaked russet with the digested shells of seemingly infinite krill. And so, during the austral spring, the penguins follow the retreating pack ice to their natal breeding sites. If the winter has been severe and the pack ice extensive, the early penguins are forced to waddle and toboggan on their bellies, single file over the pack ice, often for hundreds of kilometers. The long lines of black Adélies threading the ice are to the Antarctic what the springtime chevrons of geese are to North America or the sinuous files of wildebeest are to the African savanna. Now and then during its migration, a penguin seems to forget what it is doing; it stops in mid-track, maybe closing its eyes for a while. But soon the hormones switch on or a synapse closes, and it is on the move again.

The chinstraps arrive at Bailey Head in mid-November. They are fat from the winter's feeding and have plenty of energy reserves. The males arrive first and bellicosely declare their little territories even before the snow has melted. As the spring thaw proceeds, their real estate becomes ever more valuable. Once it reaches rock bottom, envy and greed set in. Ice-free terrain is rare in Antarctica, and penguins have limited resources for marking their nests: stones and pebbles are all there is. Coveting a neighbor's stones, filching a neighbor's stones, and fretting over one's own stones are the summer obsessions. Often the net is zero: on many occasions at Bailey Head I watched a chinstrap steal a stone from a neighbor while one of its own stones was filched on its unguarded flank, or watched a penguin fend off a potential robber and be blindsided from the rear. Soon the nest has a critical mass of stones, is comfortably situated a little more than one pecking length from its neighbors, and is accentuated by stellate streaks of guano. Like the displays and the stones, the guano is also a signal: metabolism going on here; I'm home.

A few days later, the females arrive and are enthusiastically wooed by the property-owning males. Experienced penguins seek their mates of previous years. The partnership is forged and reinforced by ritualized vocalizations and behavioral displays. During the "ecstatic" vocalization, the penguin flaps its wings, arches its neck, raises its head and, with body-wracking shudders, calls to the sky. It sounds like a cross between the bray of an ass and the honk of a goose.[9] Edward Wilson was the first to call this behavior "ecstatic," an anthropomorphism that has survived the rigor of modern-day ethology. In fact, the display is hardly ecstasy: males use it to anxiously advertise their territories and to attract females. Early in the breeding season, when land rights are still being contested, one male's ecstatic display usually precipitates a bluster of others, which spread through the colony like an acoustical wave. Later in the season, their territorial instincts aroused by the appearance of the chicks, the females join in the braying. The sound, amplified by a thousand voices, is one of the recurring acoustical motifs of Antarctica and haunts one's memories and dreams long after one leaves the continent.

Being short and piebald, with stubby, flat wings that flick erect in an instant, penguins are natural semaphores. It is no wonder that at least on land (and perhaps in the sea, as well — we just don't know), much of their communication is visual and in black and white. Ethologists have interpreted — and reinterpreted — the meanings of these displays. Each species has its own visual signature, but the general patterns are similar among the brush-tailed penguins. Adélie penguins, for example, have as many as six territorial or threat displays, four or five for courtship and mating, and some that are used for both. For a penguin, eloquence is ritual, posture, and movement. An artfully directed eye, first one side, then the other, or a silent, unblinking sideways stare, with the head downturned and the bill cocked and ready to stab, signal bellicosity in penguin language: you're too close to my nest, move on, get away from that pebble. Tucking the bill into the axilla of a wing and rolling the head in this position, accompanied by a low *grrrr,* is aggressive. These displays are much more economical than battle. But if all else fails, penguins will reluctantly, but energetically, attack: squawking, thrusting their chests and flailing their flippers, often pinning down their foe with foot and bill. I once strayed too close to a king penguin on South Georgia Island and provoked it into chasing me down the beach, punching my legs with its chest, and battering me with its sharp-edged flippers, which left bruises on my shins.

But the flippers are also the gentle tools of love. While on the nest, each member of a pair of penguins silently bows to the other like a nervous Japanese son-in-law, flippers pacifically plastered to its sides. In his precopulatory display, the male sidles up to his mate, who is coyly sitting on the nest, and wiggles his flippers. The female turns her head upward to meet his, and each fibrillates its bill at the base of the other's. Now the male climbs on top of her, and keeping her transfixed with flipper wags and bill vibrations, slowly walks down her back, straining his rear to meet hers, until finally the act is consummated by a brief kiss of their cloacas. Like all birds, the male penguin has no penis, and this dabbing with spermatozoa is as intimate as it gets.

Starting at mating time, the juvenile chinstraps run amok through the rookery. Hatched in previous seasons, they are neither physically nor socially equipped to reproduce until they are two to four years old. Regardless, they practice the behaviors that will be useful to them in the breeding seasons ahead: territorial feints, courtship displays, even stone-robbing, and they sow disorder wherever they go. The cost and risk of reproduction are high: mature penguins that do not breed are more likely to survive than those that do. In a year when food is ample, everybody takes a shot at immortality and attempts to breed, including a large number of three- to five-year-olds. Despite their physical maturity, these birds lack experience dodging skuas and sheathbills, are easily robbed of their pebbles, have not yet perfected the timing of their cycles of feeding and brooding, have immature social skills, and often fail to breed. But they learn on the job and will be more skilled the following year. During flush years the colony is densely packed and is a chaos of squabbles and bicker-ings. But in lean years, such as the recent El Niño summers of 1982–83 and 1986–87, few but the most experienced penguins risk breeding. During those summers the smallest chicks often die, outbid by their older siblings.

After two or three weeks of courtship and mating, during which time each penguin learns to recognize its mate's voice, the female lays two eggs inside the circle of stones. The timing is crucial. If the eggs are laid too late, then the chicks will not receive enough food and will lack the body mass to tough it out in the bountiful but angry winter sea. With a waddling delicacy, the female places her abdomen over the eggs and ensconces them in her brood pouch, a featherless, highly vascularized cavity that radiates incubating warmth. Because the nest of stones is a poor insulator and much heat bleeds into the ground, penguin eggs incubate at 30°C, rather than at the adult penguin's body temper-ature of 39° C.

The male immediately returns to sea, where he will feed for five to seven days. Procreation has depleted him. For a month he has fasted, all the while declaring a territory, defending it, luring a mate, and fertilizing her, and now he must fatten up on krill. Once again he is an aquatic animal, flying through the water as far as eighty kilometers from shore. When finally the sleek, gorged male returns to the nest, his mate is scruffy, caked with guano, and at her physiological limit from producing and brooding the eggs and defending them from enemies. Again, all is timing: should the male not return, or return just a few days late, she will abandon the eggs to the freezing wind and marauding skuas. She recognizes him, not by his appearance but by her memory of his voice, and both penguins begin mutual vocalizations, bellowing their identities to the sky. The female steps off the nest, leaving the eggs exposed to the wind and cold earth, and the male enters the nest, quickly transferring the eggs to his brood pouch before they freeze.

Now it is the female's turn to feed at sea. During incubation the male and female relieve each other at the nest, each time for a slightly shorter period. By the time the chicks hatch at thirty-two to thirty-five days, the shifts are occurring daily. The chicks can subsist for four days on the reserves of their yolk sac, but afterward they become insatiable and demanding. For the parents the long periods of stolid waiting during incubation are a cinch compared to chick rearing. The chicks require regurgitated krill, shelter, body warmth, and protection from skuas. The parents are on a nutritional treadmill from sea to land, converting krill to penguin flesh. By early January, one sees legions of chinstraps waddling onto the thin beach at Bailey Head, displaying to their neighbors, maybe bill-fencing a bit as they tread through the alien nests on the way to their own. Their return is greeted by the familiar cacophony of family: the mutual identity vocalizations with their mates, the frantic gaping and crying of the chicks. The chicks seem to be all stomach and gut, and from the perspective of an adult looking down at its fluffy brown offspring, they must appear to be little more than open maws. Cued by this sight, the parent bows its head and gapes. The chick forces its head halfway into its parent's gullet, contentedly gulping regurgitated krill paste. Feeding is over in seconds, and the chicks void just as quickly, squirting jets of pink guano into the demilitarized zone between the nests.

After a few weeks the nutritional needs of the growing chicks cannot be met by parents who spend half their time ashore, and the chicks, which with a bit of luck and a lot of chutzpah are now big enough to fend off skuas themselves, are left ashore while both parents forage at sea. Without the warm shelter of their parents' bodies, the chicks are vulnerable to exposure; the brown down of infancy soaks up moisture, and a freak summer shower or wet snowfall, followed by a freeze, can kill thousands. So the chicks instinctively cuddle. They form crèches: huddled groups of babies that anxiously await the return of their parents. There is safety in numbers, and crèches protect chicks in two ways: from the cold, by increasing body contact and presenting the least surface area to the elements, and from predation, by presenting skuas with a crowded and undulating mass instead of isolated and vulnerable individuals. Of course, the risk of exposure and predation is always greatest on the margin of the crèche, and those on the edge try to wedge into the warm center, pushing others to the outside. Perceived from a distance, a crèche seems to be a model of cooperation, with each animal taking its turn on the edge. In fact there is tremendous bullying for the favored spots on the interior. The bigger, first-hatched chicks invariably win, and the frontier of the crèche is littered with foundering chicks that haven't kept up with the voracious pace of summer's feeding. Some are runts that simply hatched late; others have been orphaned by leopard seals. Over the days, they slowly decline, and the difference in size and energy between them and the others widens. Toward the end, the skuas set up a vigil on the panting, flailing chicks, waiting for their little, cold deaths.

The problem of recognition among family members is made yet more difficult by the babel of the crèches. How is a returning adult to recognize its own offspring and not squander its bounty on an unrelated chick? Experiments have shown that the adults, however strongly bonded to each other's calls, do not easily recognize the voices of their own chicks. However, the chicks by now have learned to recognize their parents' vocalizations. Upon hearing an adult return from sea, they dash from the crèche — rotund, tripping tumbleweeds of brown feathers — to intercept. Sometimes three or four chicks will mob a big-gulleted adult, and in confusion it flees. But the chicks pursue, scrambling after it over the beach and through the abandoned nests. A particularly large chick can knock an adult down. Each chick is screaming, "You're my parent. Feed me!" But, of course, some are lying. There is an enormous potential for parasitism here: to cop a meal from somebody else's parent could assure survival. In the end, the most persistent chicks win and receive a ration of krill paste. The system is hardly fail-safe, but the chances are that an adult's own chicks, so fervently imprinted on the particular characteristics of its vocalizations, will persevere.

The thermal advantages of the crèche vanish on particularly hot days, when body contact becomes a liability. Once when I was at Bailey Head on a late January afternoon, the sun nudged the thermometer into the low teens centigrade. On this torrid day the chicks, their brown down absorbing the sun's energy, were nearly prostrate with heat. Over the course of the afternoon, as the sun got hotter, I sat rock-still in the shade of a lichen-painted cliff. A crèche of chicks slowly wandered my way, also seeking shade, and soon they surrounded me. One climbed onto my outstretched leg. I was background. I was rock. The chick had fluffed its feathers to release heat from its body. It was panting. I studied its yellow and black tongue, whose upper surface had retrograde serrations, an adaptation for grasping slippery krill. I studied its pale, mottled iris, which in adulthood would transform to fathomless black. This eye was able to perceive in both sea and air, winter dark and summer light, to read the semaphore signs of penguin body language. But it couldn't see me. I studied the thousands of densely packed feathers on the back of its flippers, each tiny pinna glazed with a thin oily iridescence and knitted into a svelte insulating layer. But on this day the flippers were serving an opposite purpose, as radiators of excess body heat, and their undersides were pink with swollen blood vessels. I studied the chick's pink, scaly feet, also tumid with heat-bleeding superficial blood vessels, and its long toes and black, strong nails, adapted to climbing crater walls, but on this afternoon, adroitly maintaining balance on my shifting knee.

By five weeks the brown down of infancy is replaced by the oceangoing plumage of a subadult, which lacks the full signature of sexual maturity but has the general piebald pattern of adulthood. The chicks shed their down in uneven strips and patches; some briefly look like punk rockers. The entire landscape of Bailey Head turns fuzzy. Wads of down billow out to sea, collect in the wind-shadows of rocks, and stick to the drying pools of guano. By the time the chicks are fully fledged in February and March, they have grown from 70 to over 3,000 grams. The exhausted parents return to the sea, and the abandoned chicks crowd the shoreline. The land is turning inhospitable, and they must enter the ocean too. But they know little about it, save what instinct tells them. The fledglings, buoyant and lacking in muscle tone, are easy prey for leopard seals. The seals know this and patrol offshore, bobbing their heads over the surf to watch the beach. This is their time of feast, and one commonly sees them lunging at the chicks, slitting their skin with sharp teeth, flinging them in the air, snapping them inside out, and swallowing the skinless bodies. But the chicks have a surplus of body fat and can afford to wait and lose a bit of weight. Finally, after days or even weeks of procrastination, they make the dash through the seals and disappear into the sea, not to return for at least a year.

But the penguin summer is still not over. In late March the parents and other adults return to the shore to molt their battered and stained feathers and grow the new ones necessary to survive the winter. Since feathers provide waterproofing and insulation, they cannot be replaced at sea. The penguins fast once more, dozing on the beach. It seems effortless, but feathers require protein and fat, and penguins routinely lose 40 percent of their body mass during molt. When in April they return to sea, they are once again famished. It is either feast or famine; this is the motif of their life.

✳   ✳   ✳

The brown skuas are the nemesis of the penguins at Bailey Head. This crater is theirs, and the penguins are their livelihood.[10] They perceive me as an enemy, and fly at me, low and fast and screaming, straight out of the light. Like fighter pilots, they use the sun, which is always near the horizon during the summer, as a blind, then tip into the shadow of the crater and rise again. They try to clip the top of my head with their claws. It's no feint: at two and a half kilos, adult skuas have been known to knock a grown man unconscious. I hold my hat in my fist, well above my head, as a decoy, and withdraw it just as the skuas pass overhead. It's a game that becomes tiring very quickly. Now they have learned the direction of my vision and are attacking from behind. How can the penguins be so blindly indifferent to my intrusion, and the skuas, which have no predators, be so enraged? Is it because my head projects two meters into the airspace above the penguin rookery, a three-dimensional home range that the skuas assiduously defend from all intruders? The penguins bicker over heaps of pebbles, but the skuas cut territories from the sky.

Except for the internal parasites, and the ticks, mites, lice, and occasional flea that infest their feathers and their moss-lined nests, the skuas are the apex of the food chain at Bailey Head. They are the shepherds of the penguin flock, snatching an egg or a chick when needed. Being so far removed from the blooming phytoplankton that are the source of the summer's productivity, their territorial requirements are huge: one pair of brown skuas patrols about 10,000 penguins, which in turn feed on millions of krill and incalculable diatoms. When the summer sea is bountiful, the skuas thrive, and during an El Niño year, when there is high mortality of penguin chicks and eggs, they feast.[11]

Skuas, which have the adaptive cunning of predators, are the antithesis of the robotic penguins. Skuas exploit every aspect of their environment. Brown skuas take milk from the breasts of elephant seals. They eat the scruffy alien species that humans have introduced to Antarctic stations; on Crozet Island 37 percent of skua pellets contained the remains of black rats, and on the plateau of Macquarie Island, where the brown skuas feed on rabbits, there was a 50 percent drop in the skua population after their prey were decimated by myxomatosis. Skuas even cope well with the human presence. Along with sheathbills and Dominican gulls, they have learned to forage through the garbage left by the scientific stations. Easy pickings, perhaps, but their dependency on refuse can be hazardous. Brown skuas have ruptured their crops on corncobs, have become tangled in fishing lines and aerial arrays, and have even ingested parts of lead batteries. Alien diseases are another danger; during 1978 and 1979 there was a devastating outbreak of fowl cholera (perhaps introduced by frozen poultry) among the skuas of the Antarctic Peninsula and its neighboring islands.

Today most Antarctic research stations have populations of brown skuas that are quite dependent on station personnel. Young brown skuas are easy to tame, and their tameness is culturally transmitted over generations.[12] But not all human visitors to Antarctica are trustworthy. The members of the first wintering party at the Great Wall Station of the People's Republic of China, on King George Island, impolitically killed and ate the skuas that had been so lovingly tamed by the personnel at the nearby Chilean and Russian bases.

The flightless, abundant penguins are also easy prey for the American sheathbills.[13] Pure white, squat, strong-legged, with unwebbed feet, sheathbills look more like pigeons than seabirds. In fact, although strong fliers, they spend little time at sea, and one could argue that the species is an unusual example of a terrestrial animal in Antarctica, except that most of what they feed on is ultimately of oceanic origin. They are the gleaners and garbage collectors of coastal Antarctica, eating just about everything: algae strewn on the beach, the fleshy rinds of limpet shells discarded by Dominican gulls, seal placentas, carcasses, broken eggs, even penguin feces. As an adaptation to their septic diet, sheathbills have horny shields over their nasal passages (hence their name), and pink, fleshy wattles on their faces, which may, like those of New World vultures, be rich in cleansing lymph nodes. But whenever possible, the sheathbills prefer the easy pickings of a penguin rookery. They waddle in pairs among the penguin nests, heads bobbing, pecking at scats and other morsels. They have the cunning of muggers on a city street. Working in tandem, a pair of sheathbills distracts a penguin in the act of feeding its chick, causing it to burp up a delicious little splash of krill.

Most sheathbills spend only the summer nesting season in the Antarctic; in the winter they migrate north as far as the Falklands and South America. I once investigated a sheathbill nest on the south shore of Coronation Island, in the South Orkneys. Sheathbills are solitary nesters, but since this beach was the only available habitat for kilometers around, a few pairs had built their nests in uneasy proximity to each other in the abandoned burrows of prions or beneath plates of rocks and mats of mosses that had been upheaved by frost. I heard the chicks wheezing in the rocks below, protesting my footfalls. But when I reached into one of the nests they immediately became silent: it's best for defenseless things to lie low. The nest was like Grendel's lair: a bed of penguin feathers, broken eggshells, dried egg membranes, tufts of moss, decaying food, a few bones, excrement. The three chicks wore the brown down of the newly hatched, with white tufts on the chin and just below the eyes, which may guide the parent toward the gaping mouth in the twilight nest. Probably only two of the chicks would survive to fledge; the smaller, last-hatched chick, like the superfluous offspring of the chinstraps in a bad year, was just insurance. In a place where economy means survival, it was probably destined to die and be devoured by its larger siblings.

*   *   *

By late summer the ruckus of penguin breeding at Bailey Head is over. The chinstraps head to the edge of the expanding pack ice or into the blue water beyond. The giant petrels and brown skuas also head north, but several of the south polar skuas that used to fly as far as the Caribbean now spend the winter panhandling at nearby scientific stations. The silence at Bailey Head is absolute and startling. Tufts of down accumulate under the lips of rocks and stick in the puddles of guano. The pebbles that have been shuffled and reshuffled over the course of summer, that have marched up and down the crater wall, appear to have changed not at all. Between the abandoned nests the volcanic soil is burned

5
The Galaxies
and the Plankton

The dragon-green, the luminous, the dark, the serpent-
  haunted sea,
The snow-besprinkled wine of earth, the white-and-blue
  flower foaming sea.

— JAMES ELROY FLECKER

TONIGHT THE SEA and the sky have become one. An infinite dome of cold, silent air has embraced the bay; there is not a breath of vapor, not a filament of cloud. Until long after midnight the northern horizon was etched by light, and the sea was bright and multihued, like a thin film of oil on water. The sun has reluctantly drifted north, and with no clouds to scatter and diffuse its light, darkness has descended like an opaque veil. Now there is no moon, no aurora. And for the first time since I arrived at Admiralty Bay, the sky is swarming with stars, like motes from a campfire. I feel as if I am in the center of all stellar creation rather than a remote and trivial eddy of the universe. The Milky Way is an extravagant condensate of infinite space, a clouded messenger of alien worlds distant and cold. The two Magellanic Clouds seem to hover over the bay as if they were just beyond the looming mountains, as if their light

were fresh, and not 170,000 years old and distilled of their full
spectrum by as many light-years of intergalactic and interstellar
dust.[1] Closer to Earth, waves of dense, cold air undulate across
the sky, causing the interferacting stars to now and then pop
out of the void and disappear again. Even the glaciers, gray and
dirty in the star-brightness, have paused in reverence, locked
in the silent and penetrating cold of dusk, and the ephemeral
streambeds that bleed from Flagstaff Mountain are crystalline
and stiff.

It is the night of the plankton watch, my turn to monitor the
floodlight perched on a post in the shallow sea just beyond the
floating dock in front of the station. The light is a beacon that
attracts all forms of zooplankton, as well as the predators of
those plankton, like moths to a flame. The light becomes a bright
cell of life in the dark bay. I pull on my rubber hip waders and
stride into the sea just to the depth before the cold water spills
over the brim. The waders are buoyant, endowing me with the
light gait of a moonwalker. An iceberg has stranded in the shallows, and I lay my flashlight, collecting bucket, and net on the
convenient shelf that it provides.

When I enter the field of light, the cloud of animal plankton
parts briefly, but when I stand very still the swarm returns as if I
were an inanimate object: a rock or an iceberg. There is a blizzard
of plankton under the light: copepods, hyperiid amphipods, ctenophores, mysids, an occasional squid, and krill. Most of this life
is less than a centimeter long. The copepods, armored with a huge
carapace, oar through the beam of light on five pairs of legs,
filtering one-celled algae from the sea with other, more specialized
legs. The hyperiid amphipods and mysids are predator-quick,
darting like shooting stars, sometimes looping out of control in
the pursuit of their prey. Their two black eyes are on stalks,
enabling them to see 360 degrees. The ctenophores have eight
rows of beating cillia, which oar them through the water in
iridescent paroxysms. The krill are mostly thin larvae at the stage
called furcilia, and a few adults. Except for the black pigment of
their eyes, the pulsing shadows of their hearts, and the green lines

of their diatom-stuffed guts, they are as translucent as antique glass. An icefish hides behind a rock and darts after the krill. It is so full-bellied that unswallowed krill tails stick out of its mouth. A Wilson's storm petrel is hovering near the light like a lost moth. A Weddell seal — at least, I hope it is a Weddell and not a leopard seal — is cruising in the dark sea just at the periphery of my vision, seeking the fish that are eating the krill. I hear its explosive breathing but can only dimly perceive it, a ghost in the diffuse light.

I take the flashlight and sweep its beam through the clear dark sea. The plankton cloud follows. The reasons for this fatal attraction are mysterious. Perhaps some of these organisms use bioluminescence to signal each other and to maintain the cohesion of their schools in the dark sea. Each krill, for example, has ten or twelve bioluminescent organs on its flanks. They must believe that the light is an inviting swarm, conferring on them the safety of anonymity. When I turn off the flashlight, the light-generating organs of the krill glow a cold purple-blue, looking like reflected starlight, creating their own moving constellations in the sea. I gently lower the net into the sea and switch the light back on. The krill are the most chary; they skitter around the net frame, sometimes convulsing their tails and popping backward, just out of range of capture. A few instantly shed their transparent shells, leaving them as decoys as they swim out of range. But with quick wrist work I snap the net through the beam of light, invert it and plunge it into the collecting bucket. The net has a pleasant sea-smell; its contents are a living bouillabaisse. I rub some of the amphipods and larval krill from the rim of the net and taste them. They have a nutty shrimp flavor and are slightly oily. Many Antarctic marine crustaceans of the open sea store energy as oil or wax, an adaptation that also makes them buoyant, saving them the expense of constantly swimming to keep from sinking. Some of the amphipods are so waxy that they are held by the surface tension of the water and float like husks of rice.

✳   ✳   ✳

I take the bucket of plankton to the refrigerated laboratory. The plankters swim endlessly around the bucket, actually creating a small current. I wonder how much heat, perhaps a few thousandths of a degree, their little metabolisms are releasing into the water.

I decide to first examine the slimy broth under the compound microscope, and with an eyedropper I suck a drop from the bucket and place it on a concave microscope slide. I chill my fingers in the bucket before picking up the slide. Heat is the enemy of these organisms. They can't tolerate the alien warmth of my breath or the radiance of my fingers on a microscope slide. Peering into the microscope on this cold night in the cold laboratory, I am overwhelmed by the profligacy of life, even in a single drop of water, during this Antarctic summer. At low power — a magnification of $10\times$ — chains of *Thalassiosira* dance in the light like countless stars, each scintillating, each refracting the light into a partial rainbow. These are diatoms, one-celled, glass-armored algae that are invisible to the naked eye but upon which everything, from copepods to krill to whales, ultimately depends for nourishment.[2] The diatoms swarm across the field of view, itself only a millimeter wide, like an animate Milky Way.

Diatoms are characterized by silicon shells, called tests, that are made of pectic organic materials and hydrated silicon. The silicon is in an opalescent state, refracting light exactly as an opal does. Silicon is one of the most ubiquitous of the elements, the stuff of volcanos and igneous rocks, but opalescent silicon is rare. Diatoms are ravenous scavengers of silicon. In water where the element is limited, they will remove almost all of it, reducing its concentration to less than one part per million.

I spin the objective to $100\times$, and the details appear. The pectin and silicon shells are sutured like tiny pillboxes, fretted with a radial array of pores, knobs, bumps, spines, and interstices. Their interiors are hidden by the frosted glass of the tests, but a few shapes vaguely appear as I focus through their plane: the vacuole that stores the energy-rich chrysolaminarin, an oil that is their food reserve; the brown-green plastids, the factories of photosynthesis in a realm where light is only a seasonal visitor; the grainy blur of the endoplasmic reticulum, which is the manufacturing center of the cell and which interprets the instructions of the nucleus; the nucleus itself, where the program for this mote of life is stored.

Despite their minuscule size, diatoms are not simple organisms and they are not primitive. These are advanced plants, which began star-dusting the sea relatively recently, during the early Cretaceous, 140 million years ago, at about the same time that flowering plants were beginning to mantle the land. Diatoms are sexually sophisticated, producing eggs and spermatozoa, and therefore endowing their progeny with all the splendid array of recombinant genetic permutations that sex permits. But during most of the Antarctic summer diatoms reproduce asexually by binary fission. Their tests split along the suture of the pillbox. Each daughter cell inherits half of the box and then grows another lid. Necessarily, of course, each succeeding asexual generation is smaller than the last. After several generations, this shrinking progression is stopped by the formation of an extraordinarily large cell, the auxospore, which discards both halves of the test and generates new, larger ones.[3] The progression from auxospore to auxospore takes only a few hours, and the frenzied, sun-fueled division begins anew.

*   *   *

There is only one short link in the food chain between a diatom and a one-hundred-ton blue whale — between one cell and the largest of all animals — and that link is the Antarctic krill. This may be one of the shortest and simplest food chains on Earth, but the numbers are astounding. An adult blue whale eats three to four tons of krill per day over the four-month Antarctic summer: in all, about a half billion individual krill. Each krill that perishes in the maw of the whale was wrought according to a genetic blueprint that is unique in the universe. And these half billion krill, in turn, have during their lives filtered from the sea perhaps 10 billion billion diatoms, each a mote of the sun's energy and also an evocation of self. This is far more diatoms than there are visible stars in the sky tonight. As far as we know, that number is about the same order of magnitude as the number of stars in all of the cold, invisible corners and limitless planes of the universe.

I net a single krill, three centimeters long, from the vortex in the bucket, place it in a petri dish of chilled seawater, and examine it under the low-power dissecting microscope. A few preliminary checks tell me what species it is. I note that the eye is spherical, not lobed. This tells me that it is not in the genus *Thysanoessa,* the big-eyed krills, but must be in the genus *Euphausia.* The animal has no dorsal spines on its abdominal segments, eliminating the spiny krill *(E. triacantha)* and the northern krill *(E. vallentini).* This is no surprise, since both of these species are largely subantarctic, abundant north of the Antarctic Convergence and in the coastal waters of Patagonia and the Falkland Islands, but rare in Admiralty Bay. The krill under the microscope has a stubby, blunt spine protruding from the thorax between the eyes. This detail informs me that I am peering at either an *E. superba* or a pygmy krill *(E. frigida)* and not the ice krill *(E. crystallorophias),* which is also abundant under the light tonight. The options are now only two. I examine the antennae. This krill has a bilobed lappet on the first segments of its antennae, as opposed to a triangular lappet. The whole process of keying out this animal has taken only a few seconds. It is an *E. superba.*

Most people lead their lives indifferent to *Euphausia superba,* the Antarctic krill. They don't know what they are missing — and not because someday soon their pizza may be garnished by krill. There is a greater biomass of *E. superba* — about 600 million tons — than of any other species of animal on Earth; more than humans, more than the fleets of albacore tuna that ply the Pacific, more than the locusts that periodically devour equatorial Africa. There are enough krill to sustain the indifferent predation of the baleen whales, the burgeoning population of crabeater seals, the porpoising penguins, and the darting antifreeze fish. The DNA of krill has gone berserk stamping out countless copies of itself. I am now examining a single individual of this most biomassive of all animals, one copy of unknown trillions, each sufficiently identical for me to key it to species but each, of course, unique, with its own set of recombinant genes, its own suite of serum proteins, its one particular morphometric design.

My *E. superba* is like an animate crystal refracting the hot microscope light; its clear shell scatters the beam into a partial spectrum of blues and a hint of orange. The segments — head, thorax, seven tail sections — interlock like transparent armor plating and give the body a clanking rigidity. The shell is chitinous, made of a carbohydrate related to cellulose. All of life's functions must take place within this vessel; to grow, the krill has to split its shell asunder and cast it away, replacing it with a new shell, lying nascent and unhardened beneath the old.

The thorax, which makes up a third of the krill's body length, is the largest segment. Forward, nestled in thoracic sockets, are the stalked compound eyes, brown with light-absorbing pigments. They are faceted beyond the resolution of my eye and brain. The krill pulses its thoracic legs, three pairs of short endopods that push water past its gills and three pairs of extended, bristled exopods that cup into the thoracic basket, a filter that sweeps the water of diatoms and other plankton. The filtering legs of the thoracic basket are to the krill what baleen is to a whale. Their bristles look like fibers of spun glass. Behind are six pairs of short pleopods, each pair originating from one of the abdominal segments. The seventh pair of legs emerges from the final segment and expands into the tail, spreading and contracting like the fan of a flirting geisha. The thelyctrum, the seminal receptacle at the base of the thorax, tells me that this is a female. But she is virgin. The thelyctrum bears no spermatophore, the chitin-wrapped packet of sperm that would have been placed there during a lover's brief embrace as the couple drifted among the hordes of others in the open sea. Her ovary, a smudge in the translucent thorax, is tiny and barren. This makes sense, since spawning now in Admiralty Bay, with its many predator-infested shallows, would be premature. Better to wait, to drift to the edge of Antarctica, the edge of the ecosystem, to release the depth-seeking eggs in the southward-drifting subsea currents. She has been feeding: her stomach and long gut are colored green-brown by diatoms. And her translucent heart is pulsing so fast that it appears to be fibrillating. The blood is clear and invisible.

The krill is gyring wildly in the captive ocean but getting no-where. Then, in desperation, she snaps her tail and lurches vio-lently backward until at last she becomes stuck on a strand of bright green algae. The sensory flagella at the tip of her antennae quiver slightly, tasting the water with alarm: oxygen levels are dropping, the water is warming up.

Holding her in the forceps, I insert a thin hypodermic needle between the segments of her tail and extract a drop of blood, which I place on a slide under the compound microscope. First $10\times$. Like the seawater that swarmed with diatoms, the blood eddies in its own miniature cosmos. The krill's blood cells, known as hemocytes, are spherical globs only three or four microns long that sweep past in uncountable numbers. But as I watch, constel-lations of cells distill from the torrent and freeze onto the slide. Within seconds, more cells cling to the slide and the serum becomes viscous. All movement stops, like the still frame of a motion picture. $100\times$. I can see the individual cells now, stuck to the slide. Some are granular, others are clear, having leaked their cytoplasm onto the slide in lobed blebs. Some of the hemocytes explode before my eyes, leaving a cell membrane like a punctured balloon. The krill's hemocytes are her first line of defense against pathogens. They intercept invading viruses, bacteria, and proto-zoans, rupturing and leaking their cell sap onto the invaders, clumping onto them and causing them to disintegrate. The hem-ocytes recognize other living things as foreign. This is not an easy matter on the scale of cells. They also recognize the nonliving slide, which is covered with a scum of life — the metabolic by-products of bacteria deposited from the air, grease from my fingers, cells from my breath — as foreign. By recognizing what is other, the hemocytes must also recognize self, the krill's personal genetic code. They declare each being's uniqueness in the wide sea.

I turn my attention back to the krill in the petri dish. During the minutes that I was peering at her blood, her legs have ceased pulsing, her heart has stilled. The krill is dying in the heat (a few degrees is all it takes) and her clear tissues have turned opaque, like a spreading cataract. Her proteins, adapted to subfreezing cold, are coagulating, folding, twisting, and binding onto themselves, assuming shapes that are useless to life's processes. The crystalline animal transforms into an opaque, gray corpse, like a shrimp steaming in a boil. I begin tearing her apart with dissecting needles, probing her tissues for hidden parasites. This is my science, my reason for being here on this cold, starlit night, seeking the information that this animal can only posthumously yield.

* * *

The krill in my petri dish, this trivial but unique piece of protoplasm, has led a complex and wonderful existence. She is probably no more than a year old, and with extraordinary good fortune, she could have lived up to seven years, which is long by the standards of zooplankton.[4] In her short life she has traveled through hundreds of kilometers of open sea and has traversed the vertical layers of the Southern Ocean. Krill are heavier than water and sink if they do not constantly beat their six pairs of swimming legs, a necessity that makes every one of life's activities — hunting, eating, even resting — expensive. This krill never touched the seafloor, never took refuge from the current beneath a rock nor clung to a strand of kelp. Other Antarctic zooplankton, such as copepods and amphipods, manufacture buoyant oils and waxes that keep them from sinking, but krill must constantly swim.

These wonderful peregrinations are all controlled by instinct. The fleeting electrical software that directs these behaviors is locked inside a few bundled neurons behind her stalked eyes. Yet this smudge of cells knew the waxing and waning of the pack ice and the drift of water masses the size of continents.

The Southern Ocean, pushed ever eastward and nudged into gyres and eddies by the West Wind Drift, is an unpredictable place to be a plankter. But this krill took advantage of the general patterns of ocean movement. She was probably hatched near the Antarctic Convergence, hundreds of kilometers out to sea, although she could have come from one of the smaller populations that breed close to the continent and even in Bransfield Strait. In any case, her conception took place over deep water, and the egg that was to grow into this krill immediately sank to the lightless depths, an adaptation to avoid predators in the life-filled top of the sea. A female krill may spawn as many as sixteen times during the course of the austral summer, and each batch of eggs, depending on the size and age of the female, may number as many as 2,000. Such extravagant egg production is typical of plank-tonic animals, which try to overwhelm a hungry sea with sheer profligacy.

As the egg descended into the depths of the sea, its cells divided until it had the appearance of a microscopic raspberry. By the time it reached its nadir, 2,000 meters below the surface, it was an embryo with limb buds. In this dark, cold water, at a pressure of nearly 200 atmospheres, the egg hatched into a six-legged mouthless larva, known as a nauplius. Still relying on the energy from its yolk sac, the nauplius reversed its migration and slowly journeyed back toward the sunlit surface. And as it rose, it repeatedly shed its confining shell, grew, and metamorphosed. The nauplius molted three times before transforming into a spiderlike calyptopsis larva. The calyptopsis molted three more times before transforming into yet another life stage, the furcilia, a shrimplike juvenile that for the first time resembled an adult. That transformation took place just as the krill arrived at the surface. The furcilia is the first life stage endowed with a mouth and full motility, and this krill began to feed voraciously on its fellow wanderers in the planktonic fields. The slow descent and return took from two to four weeks. Besides allowing the krill to evade predators, the deep passage served another purpose. By sinking, the egg entered the Antarctic deep water, which carried it south to the edge of the continent, and the furcilia emerged in the nutrient-rich upwelling of the Antarctic Divergence just when it was time to eat. By migrating vertically through the shearing water masses, the krill instinctively knew how to hitch rides in the currents. By rising, it traveled north; by sinking, it traveled south.

If I hadn't duped the krill with the light and caught it, the chances are overwhelming that something else would have. Only a small fraction of one percent of the eggs spawned each season survive the journey to adulthood or live to the unlikely old age of seven years. The survivors do not remain scattered over the wide sea, but are gregarious and congregate in swarms. Krill swarms are among the greatest biological phenomena on Earth. They are the planktonic equivalent of a redwood forest, of migrating monarch butterflies, or of the migration of wildebeest and zebras across the Serengeti. On March 7, 1981, a "super swarm" of krill was detected near Elephant Island by the side-looking sonar aboard the RV *Melville*.[5] The swarm had a surface area of approximately 150 square kilometers and was at least 200 meters deep. Some samples netted from the swarm contained up to 10,000 Antarctic krill (equivalent to ten kilograms) per cubic meter; yet other samples, taken minutes later and in the same spot, contained almost none. The scientists aboard the *Melville* estimated that the swarm contained between 2 and 9 billion kilograms of Antarctic krill. It also contained other zooplankton: lesser amounts of two other species of euphausids and five species of copepods, as well as amphipods and salps.

Why do krill aggregate in such astronomical numbers? The question is all the more perplexing because it is almost certain that the most important predators of krill, the baleen whales, could not feed efficiently if krill didn't swarm (nor, for that matter, could Russian trawlers economically catch krill). If krill were solitary, then whales would expend more energy seeking and catching them than they would ever derive from the krill as food.

The reasons for the aggregations become clear when one swims through a swarm of krill. The swarm is not homogeneous; instead, the krill congregate in layers, ribbons, or undulating curtains, like an aquatic aurora australis. Nor are they randomly drifting in these aggregations. Most of them are oriented in the same direction relative to one another, each probably taking cues from the subtle wakes and eddies of the others. Since krill must swim to keep from sinking, the swarm progresses in concert through the water. As they swarm, they feed, leaving behind a plankton shadow, a lifeless zone of filtered water. And this may be the most important clue of all. Like the krill themselves, the blooms of phytoplankton on which they feed are patchy in distribution, and a krill in a swarm is more likely to detect pockets of plankton than if it were solitary. The swarm, in effect, increases each krill's sensory range. As one krill darts toward the plankton, and then another and another, the signal may ripple through the swarm like a billow of wind over a field of grain. Finally, the swarming of krill, like the schooling of fish, confers a certain anonymity on each individual, which confuses predators that must hunt krill one at a time, like squids and penguins, as opposed to the mass-feeding baleen whales.[6] Swarming provides a trade-off: enhanced feeding on plankton and protection from certain predators, but increased vulnerability to others.

✳   ✳   ✳

Ecologists describe krill as the "keystone" species of the Southern Ocean. They transform diatoms into food eaten by just about every other large predator in the Southern Ocean. But along with being universally appetizing, krill are toxic because they contain high concentrations of the element fluorine, a highly reactive chemical relative of bromine and chlorine (both of which are used to disinfect drinking water and swimming pools). Fluorine is harmless in small quantities; indeed, for humans, ingesting a milligram per day helps prevent tooth cavities. But in quantities greater than ten milligrams per day, fluorine is poisonous, inhibiting enzymes, diminishing growth and fertility, and, because it concentrates in bones, deforming the skeleton. Krill scavenge fluorine from seawater (which contains about one milligram of fluorine per kilogram) and concentrate it in their chitin shells, where levels exceed 3,000 milligrams per kilogram.

Everything that eats krill ingests potentially harmful levels of fluorine. An Adélie penguin, which is about one-tenth the weight of a human, ingests about 240 milligrams of fluorine per day from the krill that it eats. How does it deal with this toxic load? One of the best strategies is simply to rapidly warm the ingested krill with body heat. When the krill die, decomposition causes the fluorine to migrate rapidly from the chitin into the digestible soft tissues; however, the enzymes that release fluorine from the cuticle are denatured at temperatures above 30° C. A penguin's internal body temperature is 38–40° C, so much of the fluorine remains in the indigestible chitin and is excreted in the feces. Most birds, including ducks and chickens, have gastric ceca that enable them to digest cellulose (and its chemical relative chitin). But penguins lack ceca and pass the chitin undigested through their gut. Also, it takes only three to four hours for a krill shell to pass through an Adélie's gut, minimizing the potential for absorption of fluorine. Yet even these adaptations aren't enough, and penguins do absorb high levels of fluorine, which is sequestered in the bones until it can be excreted by the kidneys.

Penguins, of course, are not the only krill-eating animals that have to cope with high levels of fluorine. Other birds and mammals are also able to denature enzymes by body heat and store fluorine in bone. However, cold-blooded animals, such as fish and invertebrates, lack the advantage of elevated body heat (on the other hand, they do not have to stoke their internal fires with large amounts of food). And invertebrates, such as squid, lack even the advantages of bone. Their mechanisms of dealing with excess fluorine are largely unknown.

Krill fluorine is a natural biological marker. By measuring fluorine levels in the tissues of Antarctic marine invertebrates, biologists are able to decipher the patterns of Antarctic food chains. Octopuses and sea spiders have low levels of fluorine in their tissues, indicating that they survive independently of food chains based on krill. But isopods, amphipods, sea stars, and brittle stars all have high concentrations of fluorine in their bodies, indicating that they are directly or indirectly dependent on krill.

And now humans, having to feed their ever-increasing population, are turning to the sea, and to krill. Antarctic krill meat, which is 10 to 11 percent protein, has been added to the daily bread of Russian workers. It has been touted as the protein panacea for the starving people of sub-Saharan Africa. It adorns Japanese rice crackers. Advocates of harvesting krill have claimed that the estimated 70 million tons once annually consumed by the blue, fin, and sei whales that were hunted to near extinction in this century, are an unused resource that is up for grabs. Some studies, particularly those by Soviet scientists, have indicated that the huge swarms of krill in the vicinity of South Georgia are postreproductive and are doomed to die as they are carried by currents to the warmer waters of the north. The exploitation of these krill, the researchers claim, would have no effect on the reproductive rate of the species.[7] From 1976 to 1988, Soviet trawlers annually trawled up to 454 million kilograms of krill from these waters and the northern Weddell Sea.

But at least for the time being, the high fluorine content of krill has saved the species from massive exploitation by humans. Krill are usually frozen with their shells on, but even in the frozen state, the enzyme-driven migration of fluorine from shell to meat takes place at full speed. One solution, of course, is to mechanically shell the krill and save only the meat, but this is a labor-intensive activity that is impractical aboard a ship in the Antarctic. Experiments are now under way in Russia to chemically separate krill muscle proteins from the shell, yielding a low-fluorine paste that has been used to replace 30 to 50 percent of the meat in sausages.

The central role of krill in Antarctic food chains is unlike that of any other species, and a krill harvest would be unlike any other fishery on Earth. Most marine animals taken by humans are predators high on the food chain, and removing them has only a slight harmful effect on the lower levels. Indeed, it may actually foster a population increase of the prey species, as occurred among the sprats and anchovies of the Baltic and Caspian seas after the commercially important large predatory fish were removed. But a krill fishery concentrates on the next-to-lowest link of the food chain, and all higher echelons are affected. Important questions must be asked. Will the human exploitation of Antarctic krill prevent the recovery of the blue, fin, sei, and humpback whales? After all, the 70-million-ton "surplus" of krill that humans are eyeing was once consumed by these whales, and a large-scale krill harvest in the Antarctic would basically shift fishing activities from the whales to whale food.

In fact, it may already be too late for the great whales to recover their former numbers, regardless of our appetite for krill. It is likely that the "surplus" has already been consumed by the expanding populations of other krill-eating species that were not hunted by humans, such as penguins and crabeater seals. The populations of krill-eating penguins have soared in the decades since whaling ended; there has been a 3 to 4 percent annual increase in Adélies and a 7 to 10 percent annual increase of chinstraps at Signy Island in recent years. The crabeater seal, estimated to have a population of 10 million, is now the most abundant marine mammal on Earth. The little minke whale, the smallest of the Antarctic baleen whales and therefore the last to be hunted by humans, actually expanded its population during the decades of whaling, owing to decreasing competition from its larger brethren. Competition for krill from the burgeoning populations of seals and penguins may prevent the populations of large baleen whales from recovering. The keystone remains; only its equilibrium has shifted. And if humans further alter the balance by becoming a major predator of krill, then the chances of ever seeing the Southern Ocean filled once again with great whales become even more remote.

✳   ✳   ✳

During some winters Admiralty Bay never freezes. But in most years the bay glazes over with pack ice, and during the coldest winters the pack ice extends far beyond the tip of the Antarctic Peninsula and a hundred kilometers into the Drake Passage. The freezing of the Southern Ocean is the greatest seasonal event to take place on Earth. Approximately 16 million square kilometers of pack ice — one third of the area of the continent — appear and disappear every year, and in the heart of the austral winter, about 8 percent of the Southern Hemisphere is covered by sea ice. The whole world changes when the Southern Ocean freezes. Sea ice dampens waves, reflects the sun's energy back into space, and creates downward-flowing ocean currents of dense, salty water. The patterns of ocean circulation shift, and the atmospheric convection currents that drive Earth's climate alter their direction and momentum.

This change isn't a sudden congelation but a gradual advance of freeze and thaw, freeze and thaw. The freeze begins on the first still, cold night, when the bay seems oily and glazed; the minute crystals of ice stiffen the surface tension of the water. This is an audible process: the water pings and snaps almost below the threshold of hearing. The sea, losing its heat to the air, becomes veiled in fog. The first freeze usually dissipates with the morning sun. But sometimes it stays, and the wave-shuffling of the ice fibers causes them to coalesce into round plates with raised edges, their own miniature pressure ridges. These plates are called "pancake ice," and the bay looks blotched and mottled. Usually the pancake ice doesn't last very long. The first strong wind pushes it into the shore, where it collapses into undulating slush. But when the cold deepens a few nights or weeks later, the pancakes fuse and form the pack.

The ice insulates the sea from the lunar-cold air, keeping it liquid and relatively warm all winter long. The frozen sea surface may appear wind-smoothed and snowy, perhaps rilled by occasional pressure ridges and jumbled ice, but the underside of the pack is rotten and jagged. As sea ice congeals, pockets of brine form within it, and these blebs of unfreezable water migrate through the ice, like vacuoles in a paramecium, until they are expelled into the sea. Pressure ridges, formed when plates of ice collide, hang like inverted rubble. Beneath the new pack ice of autumn, the sea remains illuminated, and if you swim under the ice like a seal, the leads and polynyas beckon of air and light. The penguins and seals, and even small whales like minkes, stay in these leads. As the sun wanders north and snow covers the pack, the light wanes. Smothered by the pack and sheltered from the wind, the still sea loses its sediments and becomes transparent. Light, if there were any, would travel 80 meters before being absorbed.

Some phytoplankton get locked in the ice as it forms and become concentrated in the brine channels along with the sea salt. The density of ice algae can be enormous: one survey counted 33 million algal cells per liter of ice. There it remains dormant, the seed of the next spring's thaw and the next summer's bloom. At least sixty species of Antarctic diatoms are able to withstand freezing in the pack ice, and many other plankton species, including at least eleven more species of diatoms and several species of dinoflagellates, continue to photosynthesize in the water under the ice. Aggregations of phytoplankton hang down into the water like brown streamers from the underside of the floes. These algae are shade-tolerant; even on a sunny day, the underside of the ice receives only about one percent of the surface light, and from late autumn to early spring, the light levels become vanishingly small.

This is the *sympagic* biome: life within ice, life coating the underside of ice. I once sailed into Foster Harbor, on Deception Island, one clear spring day when its margins were still locked in ice. The pack ice was rotten and easily cleaved by the ship, and sometimes the plates of ice would flop over in our wake. The undersides were brown. At first I thought it was sediment, or paint from the hull, but on closer examination I realized that the discoloration came from billions of diatoms, each distilling energy from the sunlight and depositing it in one-cell edible packets on the underside of the ice. Joseph Hooker was the first to describe the sympagic algae. "The waters and the Ice to the South Polar Ocean," Hooker wrote, "were alike found to abound with microscopic vegetables. . . . They occurred in such countless myriads, as to stain the Berg and Pack-Ice, . . . they imparted the Brash and Pancake-Ice a pale ochreous colour."[8] But only in the 1970s was the full importance and extent of the sympagic realm fully appreciated, in both the Arctic and the Antarctic. It may contribute as much as 12 percent of the total primary productivity of the ice-bordered zones of the Southern Ocean.

It was once thought that during the winter krill drifted deep and remained dormant. But recently, remote television cameras peering under the winter ice have showed the bottom of the pack ice dancing with krill, which brush their thoracic baskets through the border zone of liquid and solid, mowing the ice algae in strips like a farmer cutting hay. Now and then an exhausted krill catches a bubble of air in its thoracic basket, and bobs upward against the ice. At last the krill, which must spend most of its life swimming to keep from sinking, has a bottom on which to rest, but it is a topsy-turvy one.

During the austral spring, the pack ice retreats. Like its formation, the breakup is not a sudden process. The pack fractures, and then the wind shifts and the sutures close once more. Some ice, carried south by the gyres of the Weddell and Ross seas, survives several years before melting. But most of the pack ice melts during its first summer, breaking into ever smaller pieces, until they are only several meters wide. Invisible to a ship's radar, these are known as "growlers," because of the sound they make as they rattle along the hull of a ship. The dissipating pack ice drifts with the sea, but the tall icebergs, pushed by the wind like sails or nudged by unseen currents, often plow through the pack, leaving behind a trail of open water.

As the ice melts, the frozen algal cells are released from their dormancy to sow the ocean with the seeds of the spring bloom. The bloom follows the retreating ice right into Admiralty Bay. As spring advances, the growth rate of phytoplankton along the ice's retreating edge doubles every day, climaxing in late spring or early summer. It is quickly followed by a burst of zooplankton production — the krill, copepods, and amphipods that graze on the planktonic algae — and eventually, of course, by the whales, seals, penguins, terns, and petrels that feed on the krill.

About 40 percent of the energy necessary to melt the pack comes from the upwelling of deep, warm water currents from north of the Antarctic Convergence. The sun provides the rest. By absorbing the energy of the returning sun, diatomin, the brown photosynthetic pigment of the ice-locked diatoms, accelerates the spring thaw. The full influence of the algal pigment is unknown, but it is clear that by subtly warming their immediate environment, these one-celled algae alter global weather patterns thousands of kilometers away. They alter the trajectories of ocean currents and the exportation of Antarctic cold to the lower latitudes. Indirectly but inexorably, these algae may affect the crops of soybeans in southern Brazil, the anchovy harvest of Peru, and the dry winds over the Sahel.

6

The Bottom of the Bottom of the World

Roll on, thou deep and dark blue Ocean — roll!
Ten thousand fleets sweep over thee in vain;
Man marks the earth with ruin — his control
Stops with the shore.

— LORD BYRON

IT IS SNOW-HAILING this morning when I make my first scuba dive into Admiralty Bay. At dawn the bay was so still and so profoundly blue that sea and sky seemed to merge on the horizon. The bay grew snow-capped mountains and the sky became liquid. But now, a few hours later, a front of dirty clouds has passed over King George Island. No doubt it is the long spiral finger of a distant cyclonic storm. But no matter; the front brought no wind and the bay remains tranquil. It is a wonderful morning for a swim.

Scuba diving in the Antarctic is the closest thing on Earth to walking in space. Today the water temperature in Admiralty Bay is 0.4° C, so cold that an unprotected diver would lose consciousness and die after only a few minutes' exposure. I don several layers of woolen underwear and socks, a one-piece neoprene dry suit, a hood, booties, and three-fingered mittens. The suits are hung from the rafters of the generator room to dry. But my suit is not dry: its interior is still damp and slightly fetid from the perspiration of the previous user. Sitting on the bench, I slide my legs inside. Then my arms and head. The suit is so snug that I practically strangle just pushing my head through the neck cuff. I walk as stiff-legged as an insect. The hood muffles my hearing and I unconsciously shout like a partially deaf man. All of my body heat is reflected back at me, and it is a sauna inside; my armpits become damp, and my forehead, which is not covered by the rubber hood, drips with perspiration. Immediately I understand the discomfort that a seal or a penguin, insulated by blubber, must feel on a warm day.

I waddle to the beach and sit on a whale vertebra to prepare the rest of the equipment. The mask, with a built-in regulator, covers my entire face. This is a safety feature: if the suit tears and I lose consciousness, then the regulator cannot drop out of my mouth. In case the primary regulator fails or freezes, a spare mask dangles from my side, always within reach. But although I have rehearsed changing masks many times in warmer waters, I can't imagine the disorienting shock of doing it in this cold sea. Setting up the equipment on the beach is an ordeal. The metal tank and regulators are too cold to touch comfortably with my bare hands, yet I can't operate them while wearing the bulky, three-fingered gloves. Edson, my diving buddy this morning, holds the tank while I slip into its backpack. The tank weighs thirty kilograms and I must hunch forward in order not to fall backward. Edson connects a low-pressure hose to the dry suit and inflates it. On the chest of the suit is a control panel with buttons for inflation and the release of air. A skilled diver, with weights properly balanced, can operate these like the buttons on an elevator. To compensate for the buoyancy of the air, I strap on sixteen kilograms of lead weights in a belt around my waist and attach two and a half kilograms to each of my ankles. Without the ankle weights, the air in the suit tends to enter the legs, causing the diver to bob head-down in the water. Unable to right themselves, divers have drowned this way. I rub Vaseline on the seals around my wrists, ankles, and neck, and put on the gloves, flippers, and mask. Edson checks my equipment a final time, and I stagger into the water. Edson will not accompany me on the dive but will wait on shore. A tether is attached to the air tank, and I drape it through the palm of my right hand. Edson will periodically tug on it to signal me; if I fail to respond, then he will pull me to shore.

The blessed cool and the weightlessness are an enormous relief as soon as I enter the water. It actually feels good to be in this sea that is so cold it would be excruciating to bare skin. But when I immerse my head in the water, the little band of exposed forehead above the mask gives me a momentary blinding headache, and the seals around my wrists, ankles, and neck begin to leak slowly. This is a temporary discomfort; soon the woolen underwear sops up this water and warms it to close to body temperature.

Today I will descend to about fifteen meters to collect the tiny amphipods that skitter over the ocean floor and hide beneath the boulders that have been dropped by the icebergs passing overhead. Since I am studying the pathologies of these crustacean relatives of shrimps, I want representative samples of both diseased and healthy animals. A baited trap would attract only those amphipods with strong appetites, which are necessarily the healthy ones. A trawl net or bottom grabber would favor the slow and sick animals that couldn't escape, biasing the sample toward the unhealthy ones. So I use a simple hand-held net, purchased in a tropical fish store in New York. The fifty kilograms of diving equipment, worth thousands of dollars, is all configured simply to bring this simple net within reach of these tiny prey. Only a few dozen people have gone before me into this realm. Until the late 1950s, when humans first began scuba diving in Antarctic waters, this seafloor was beyond the reach — and the ken — of all but the most adventurous explorers. Even now, we visit only the shallowest and most hospitable areas of the Antarctic continental shelf. Most of this dark and cold ocean floor has never been witnessed, is unknown.

On this still morning, underwater visibility is twenty to thirty meters. Flocculent streamers of brown organic matter, a mixture of diatoms and penguin guano, hang from the surface of the water. Even through the muffling hood and above the clattering of the regulator, I hear a sea filled with popping and snapping sounds. From what? Crustaceans? Perhaps. But most of the sounds seem to be coming from two icebergs stranded on the seafloor ahead. I swim next to one of them. It is releasing puffs of fine, silty sediment, scraped from the base of Stenhouse Glacier, into the clear water. As it melts in the summer sea the iceberg emits columns of bubbles. I steady myself by placing a hand on a small berg, and it rolls over, almost hitting me on the head. Edson tugs in alarm, but I signal back that all is well. An important lesson for diving in the Antarctic: never assume that an iceberg is stable.

Ice is the dominant environmental factor of the Antarctic shallow-water marine communities. Not only does it abrade the ocean floor, scraping away benthic organisms and mixing layers of sediment, it also affects salinity, temperature, currents, and the amount of light that penetrates the sea. The seafloor beneath me is a plane of glacial silt and sand, punctuated by occasional boulders. Icebergs have plowed furrows in the soft bottom. The boulders have been worn smooth by their long, grinding passage beneath Stenhouse Glacier. Some are covered with a fur of brown diatoms or green filamentous algae; others are scraped clean. Most of the rocks have one or two Antarctic limpets grazing on them.

The Antarctic limpet is one of the few benthic organisms that is able to cope with the caprices of scraping ice. The limpets retreat to the safety of cracks and fissures at the slightest hint of ice bumping their rock. This is an elaboration of characteristic limpet behavior, for all limpets, even tropical ones, are well known for their territoriality, returning to the same spot each morning after a night of grazing. The limpets also migrate according to the seasons. In the winter, when the sea is considerably warmer than the air, they hie to deeper water and wait out the cold. But during the summer they slide into the intertidal zone, and for brief periods even exceed the high-tide mark. Coated with a mucus that inhibits the formation of ice crystals, they can withstand exposure to freezing air. Their payoff is the abundant algae and lichens that coat the intertidal rocks, which are exploited by no other grazing animal. But the exposed limpets are easy prey for Dominican gulls, which pluck them from the rocks and swallow them whole.

As I swim over the seafloor, a flurry of tiny amphipods belonging to various species are also grazing on the algael beds growing on the rocks. Some are *Bovallia gigantea,* despite their name only two or three centimeters long, which stand poised on their rear four pairs of legs, tail tucked underneath the abdomen, head elevated, and antennae brushing the water. This is their characteristic alert posture. A stark silhouette against the bright light from above, I must appear to be a seal or penguin. I release a few bubbles of air from my suit and slide down onto the soft plain, stirring up a puff of sediment. The amphipods scurry beneath a rock. I nudge the rock aside and sweep the net in the depression it leaves. The net quickly fills with the little creatures; turning it inside out, I place the amphipods in a transparent Ziploc bag. Alarmed at being lifted into open water, they dance on the bottom of the bag, trying to return to the seafloor.

My quarry today is *Pontogeniella brevicornis,* one of the most common amphipods in Admiralty Bay. Like *Bovallia* and indeed most Antarctic marine invertebrates, the species has no colloquial name. The continent has never had a race of native people to name its creatures. And, as is typical of all but the most conspicuous Antarctic marine invertebrates, the species was discovered and formally described by scientists only in this century. At least thirty-three species of amphipods live in Admiralty Bay. Most are necrophagous — eaters of dead bodies — a common niche in the lightless, deep Antarctic waters where there is no primary productivity, but where there is a rain of organic material from the sunlit waters above. But *P. brevicornis* feeds on the bounty of the shallow banks, which during the summer are flooded by sunlight and coated with algae.

Releasing more air from the suit, I ease down the undulating mud slope to a depth of ten meters. At this depth, beyond the reach of the gouging pack ice, the Antarctic continental shelf changes its appearance. The soft bottom gradually loses its ice scars and becomes a plain etched by the tracks of foraging animals. The deepest furrows are made by asymmetrical sea urchins. Covered with short, soft spines, they look like calcium mice. At first the spines seem to be rising and rotating in confusion, but in fact all are netted to the same simple nervous system and working toward a common purpose: pushing the urchin through the edible mud like a bulldozer. An armored, animate gut: what wonderful economy in a realm where resources are few!

Brittle stars, long-legged relatives of the urchins, are also pressing their mouths to the mud, searching for buried clams. In some areas they are so abundant that they are heaped on top of each other. In spite of the clanking calcareous plates of their external skeletons, they are lithe, pulling themselves over the mud in short spasms. Each spasm stamps a partial rosette in the mud, and there are thousands of these images. A few of the brittle stars delicately stand on their tapered leg-tips, a common behavior for which there is no obvious explanation. Fish hide in the cupolas formed by their legs. In an abyssal plain where every feature is rapidly smothered by mud and where there are almost no hiding places, the fish make do with what shelter they can find. Watching me nervously, they hold their ground until I approach to within about a meter, then dart onto the dark plain ahead.

A close inspection of the seafloor reveals that it is densely pocked by the siphons of embedded clams, the prey of the brittle stars. With my mitten, I fan away the silt, revealing vast numbers of *Mysella charcoti*, the most abundant bivalve in the shallows of

Admiralty Bay. From the safety of their burrows, beyond the reach of the prowling sea stars, *M. charcoti* filter organic material from the water through their siphons. Their numbers are astounding; in the South Orkney Islands, the species has been collected in densities as high as 75,000 individuals per square meter of ocean bottom.

The gloomy plain ahead invites me to swim deeper, to fifteen meters. I am now beyond the reach of even the heaviest winter pack ice. Unlike the shallow sea, this is a realm of relative stability: of near constant temperature and a steady rain of silt and nutrients from above. Here nemerteans — meter-long wormlike organisms — lie on the surface of the mud. Each is striated by hundreds of annulations, the tracings of the muscles that push it peristaltically through the ooze. A wad of them, the size of a basketball, congregate in a prolonged communal copulation. Nemerteans are gregarious at the supper table, as well. Hundreds of them will heap atop a dead seal or other carrion on this sterile plain. But they are also hunters, equipped with a ten-centimeter-long proboscis that shoots a dart into their prey. On occasion they have been known to overpower and eat fish.

At these depths the ooze is alive with *Serolis polita,* a flat isopod the size of a thumbnail. Although abyssal species of *Serolis* are found as far north as the Bahamas, the genus probably originated in Antarctica. Their flat, segmented carapace, arched legs, and bilobed compound eyes, each mounted on a high turret, give them the appearance of extinct trilobites. Every *Serolis* is an entire ecosystem unto itself — an island of hard substrate in a seeping, ever-shifting bottom — and is colonized by bryozoans, algae, fungi the shape of white stars, and hydrozoans that look like red chalices. *Serolis* passes its life in cold slow motion and sheds its shell only once or twice a year. Its encrusting organisms, however, are denied the luxury of slow development and must complete their life cycles according to the schedule set by their host. In the course of a few months, they colonize, grow, and reproduce before being sloughed off with the shell.

The female *Serolis* are swollen with bright pink eggs, and when they swim, oared legs beating in tandem, they look like flower petals in the dark water. The eggs, which number about forty, are tucked into a ventral pouch, where they are brooded for twenty months. During that time there is virtually no egg mortality. *Serolis* is typical of Antarctic benthic marine invertebrates, most of which produce a relatively small number of large, yolky eggs and nurture them to an advanced state of maturity before releasing them into the environment — an adaptation that economizes on energy in a realm where predation is high and food is limited. These conditions lead to nurturing, to parental care. The planktonic strategy of casting one's spawn to the currents won't do for the seafloor of Antarctica. This pattern is typical of Antarctic marine mollusks, polychaete worms, annelid worms, crustaceans, and sea stars. Juvenile *Antarcturus,* a common isopod in Admiralty Bay, attach themselves to their mother's antennae and hang like Christmas ornaments. Even Antarctic sponges tend either to be viviparous or to have some other form of brood protection. In these cold, starved waters, eggs are expensive to produce, and by retaining them the parent increases the chances that they won't be carried beyond the narrow continental shelf by the strong currents of the Southern Ocean. Indeed, only seven Antarctic benthic marine invertebrates are known to have planktonic larval stages. One is the Antarctic limpet; the rest are sea stars and their relatives. Not surprisingly, all seven occur in great abundance throughout the Southern Ocean. Even so, these broadcast spawners produce relatively few eggs compared to their relatives in warmer waters. Brooding of eggs and larvae, combined with the scarcity of island stepping stones in the Southern Ocean, has for millennia isolated the Antarctic shallow-water marine invertebrates from the rest of the world and has led to high levels of endemism: 90 percent of the species of amphipods and sea spiders, 77 percent of the sea urchins, 66 percent of the isopods are restricted to these waters.

Slow rates of development and growth and concomitant longevity are typical of Antarctic marine invertebrates. *Serolis polita* doesn't reach sexual maturity until it is three years old. *Bovallia gigantea* males take eighteen months to mature; females take forty to forty-two months. An Antarctic limpet, in the unlikely event that it is not snatched by a gull or eaten by a sea star, has the potential to live for a hundred years. Some species of Antarctic sponges may survive several hundred years in the still and unchanging depths of the continental shelf.

After thirty minutes I have now arrived at the farthest point of my dive: a ridge of rock that rises from the floor of Admiralty Bay. A hard bottom makes all the difference. Unlike the unconsolidated silt, it is festooned with living things. This rock is valuable real estate, a place from which sponges, tunicates, soft and hard corals, and mosslike bryozoans can project their siphons, mouths, and tentacles into the nutritious water. These fixed organisms in turn provide habitat for all manner of invertebrates and fish. At this depth, the full spectrum of sunlight is filtered by the water and ice above, and the seafloor appears gray and gloomy. But when I turn on my light, the beam is like a paintbrush: everywhere I sweep it I see splashes of color. The ten-centimeter-diameter anemones, which basically consist of a mouth (that doubles as an anus) fringed by tentacles, are orange. Five-centimeter-long white brachiopods, attached to the seafloor by red stalks, reach into the nutrient-bearing water. The two halves of their shells have an overbite, and are striated with hundreds of growth rings. The drooping soft corals, which look like long catkins, are bright yellow. Each is a colony of thousands of tentacled polyps embedded in a flexible protein matrix. Each polyp, although only a fraction of a centimeter wide, is morphologically identical to an anemone. And each, like an anemone, presents its mouth to the sea.

There are even a few foliating red and green algae at this depth, and aquatic lichens paint the rock face a deep purple. Some of the rock itself is yellow calcareous algae, which extracts dissolved calcium from the water and combines it with carbon dioxide to form calcium carbonate. These plants barely survive here; like the mosses and bundle grasses that grow beneath the cover of spring-time ice and snow, they must achieve photosynthesis in diminished, filtered light. They scavenge light from the sea, just as the animals with which they share this rock scavenge organic material. Red sea urchins browse on the algae, prising it from the rocks with a rosette of spiny teeth. Bristled polychaete worms insinuate themselves into this living terrain, sometimes entering the siphons of the sponges and tunicates themselves. They browse the bryozoans and may munch on the tissues of the very hosts that shelter them. Frilled nudibranchs — snails with internal, rather than external, shells — scrape the algae off the rocks and feed on the polyps of anemones and corals.

A giant Antarctic isopod, ten centimeters long, walks through the rock garden on six pairs of legs. Segmented, its shell dorsally keeled with flared margins, it looks as if it has been chiseled from rock. James Eights was the first biologist to describe this animal. He aptly named it *Glyptonotus,* meaning "sculpted back," and wrote, "This beautiful crustacean furnishes to us another close approximation to the long lost family of the *Trilobite.*"[1] Eights found all of his specimens of *Glyptonotus* washed up on the beach after a storm. Indeed, in Eights's time most Antarctic marine biology was the result of beachcombing. Eights never had the privilege, as I do now, of observing the giant isopod in its natural realm. Isopods this size are most unusual in shallow waters, but they are abundant in the limiting conditions of the abyssal depths, where, as on the Antarctic continental shelf, slow growth to a large size is a common adaptation to the scarcity of nutrients.

Another giant, an eight-legged sea spider with about a twenty-centimeter leg span, is standing on pointed legs in the ooze at the base of the promontory. Its rust-colored skin is crinkled by colonies of hundreds of bryozoans. The "sea spiders" are misnomered; despite a superficial resemblance to spiders, they are only distantly related to them. They are in fact unlike any other group of arthropod animals and comprise their own subphylum, the Pycnogonida. The pycnogonids seem to be all legs; what little body they have consists of a head, a short neck (bearing four eyes on a tubercle), and six cylindrical posterior segments. These segments are so reduced that parts of the gut and the female reproductive tract extend into the legs. This pycnogonid seems to move in hesitant slow motion on its tiptoes. The current from my flippers is sufficient to tip it over, and it begins a labored process of righting itself. These animals aren't built for the chase, but neither are their prey, sedentary organisms such as sponges, corals, and bryozoans. All pycnogonids have a large, extended mouth known as a proboscis. The proboscis is so massive and complex, relative to the minuscule body, that it is endowed with its own ganglion of nerves — in effect, a secondary brain. Many species use the proboscis to suck up the tissues and fluids of their prey in a prolonged lethal kiss, and some also have paired fangs, known as chelicerae, which are used to prise off tissue and pass it to the mouth. This pycnogonid has a long, blunt proboscis designed to probe the soft glacial sediments. Having righted itself from my flipper gust, its strategy is just to stand there. It is a formidable animal, as big as my palm, another example of gigantism in the Southern Ocean.

Most pycnogonids are smaller than a fingernail and, although common throughout the world's seas, are seldom noticed. The sea surrounding Antarctica is denizened with approximately a hundred species of pycnogonid, several of which are giant. Most, like the one I am watching, have eight legs. But some have ten legs, and a few have twelve. James Eights was the first to describe a ten-legged pycnogonid, in the genus *Decolopoda*, from a specimen with a fifteen-centimeter leg span that he found among the flotsam on a beach in the South Shetland Islands. It was but a variation on the theme of the one I am watching, with a shorter proboscis designed to suck bryozoans and other encrusting organisms from the faces of rocks and the blades of algae. Eights wrote, "They are found in considerable numbers in connection with the fuci, thrown up by the waves along the shores of the islands, after being detached by the motion of the large masses of ice, from the bottom of the sea."[2] Until the genus was collected again in 1905, specialists considered the ten-legged specimens to be anomalous individuals of eight-legged species. In 1931 a twelve-legged pycnogonid, *Dodecolopoda mawsoni,* with a leg span of seventy-five centimeters, was collected in water 219 meters deep off the Mawson Coast, in the Indian Ocean sector of Antarctica. It was, and remains, the largest pycnogonid ever collected. Since then, two more specimens have been collected in the Ross Sea, and a second twelve-legged species, *Sexanymphon mirabilis,* has been taken from the waters of the South Shetland Islands.

The sponges on this promontory are mostly white and tan; a few are dirty yellow. Some are encrusting and adopt the shape of the rock on which they are growing. Their smooth surfaces are pocked with the incurrent and excurrent siphons through which they filter plankton and organic material from the seawater. Others stand freely, cupping the water and catching the nutrients from above. Sponges are colonies of cells that are barely consolidated as organisms; they have no discrete organs, and the division of labor among their cells is rudimentary. Yet this lack of cell specialization endows the sponges with great powers of regeneration. One can mash the tissues of a living sponge through a coarse cloth, and the resulting inchoate broth of cells will reassemble into an entire new organism. Like the soft corals, the sponges have flexible internal skeletons, made of interlocking pieces called spicules. Most sponges in shallow tropical waters have spicules of flexible protein, but in the deep sea and on the Antarctic continental shelf, the dominant group is the glass sponges, whose spicules are made of brittle silicon. Like the diatoms on which they feed, they weave their bodies from the wine of the land.

The tunicates, nacreous and as pale as a dead man's fingers, also seem part of the fabric of the rock.[3] As adults, they have the general appearance of sponges, with a tough protein and cellulose outer tunic (after which they were named) and single incurrent and excurrent siphons through which they filter the water. But a tunicate is not a barely organized chaos of cells, as is a sponge. A cosmos of evolution separates them from the sponges. Tunicates have a gut, nervous system, and circulatory system, complete with a heart. More remarkable, their larvae resemble tadpoles, complete with a dorsal nerve cord and a rudimentary backbone called a notocord. This trait places them in the Chordata, the same phylum as fish, amphibians, reptiles, birds, and mammals. This trait places them in my phylum.

Serolis and its giant isopod relatives, the giant sea spiders and the glass sponges, so conspicuous and abundant in the shallow fringes of Admiralty Bay, are characteristic forms not of the shallow sea but of the deep benthos, the abyssal plain kilometers beneath the surface. In terms of the fauna, it is almost as if I were diving in the Marianas Trench, viewing scenes through the thirty-centimeter-thick porthole of a deep-diving submersible. The fauna of north polar seas is similar.[4] Zoologists have long recognized that families and genera of certain deep-sea invertebrates are also found in the shallows of the polar areas. The reasons are obvious: both ecosystems have constant cold temperatures and must depend on nutrients from above.

Amid all of this diversity, certain groups that are ubiquitous in the other seas of the planet are missing: decapod crustaceans, which include crabs and shrimp, and the cirripedians, the familiar barnacles that adorn rock faces and pylons in all the oceans of the world except the Antarctic. It is easy to understand why barnacles fare poorly in Antarctica. As adults they are filter-feeding animals that are permanently cemented to an object, such as a rock or piece of wood, and are necessarily confined to shallow, sunlit waters where plankton abound. This is exactly the zone that is scraped by ice in the Southern Ocean. The absence of crabs, prawns, and shrimp is more of a mystery. Only three species have been described from Antarctica. Two of these, although rare, seem to be circumpolar in their distribution. The third, *Lebbeus antarcticus,* has seldom been observed. It was first described in 1941 from a single specimen. Since then only twelve more specimens have been captured, eight of which were taken, partially digested, from the stomach contents of two female Weddell seals in the southern Antarctic Peninsula. Why have crabs and shrimp, so eminently successful elsewhere, not made it in Antarctica? One reason is the isolation of the continent by the Southern Ocean and the West Wind Drift. Crabs, prawns, and shrimp are shallow-water organisms that tend to broadcast an abundance of eggs and larvae into the planktonic swarm. Many shallow-water colonists simply couldn't survive the ocean crossing, either as larvae or as adults. However, their close relatives (and therefore strong competitors) the isopods and amphipods, which crept along the ocean floor and emerged on the Antarctic continental shelf, found the crossing less perilous. Many abyssal isopods and amphipods also brood their young and were therefore preadapted to Antarctic conditions, making them strong competitors against shrimps and crabs.

I can't count the species of invertebrates that have colonized this rocky outcrop and muddy plain, but there must be hundreds. To date, taxonomists have identified 129 species of tunicates in the Antarctic, 875 species of mollusks, 650 polychaetes, 470 amphipods, 310 bryozoans, 299 isopods, and 100 pycnogonids. Probably as many more species are as yet undiscovered. But the proliferation of sponges may be the most surprising of all. More than 300 species have been identified in Antarctic waters. In some areas they are the dominant fauna, and the seafloor is a pavement of silicon sponge spicules, the result of millions of sponges slowly growing and then dying over countless thousands of years in the dark sea. Antarctica may be the world center of sponge evolution, with more living species than any place on Earth. Indeed, Antarctica today may have the greatest diversity of sponges of any place at any time in Earth's history, more, even, than the Paleozoic seas before the appearance of the vertebrates.

This is the revelation that greets the few who have journeyed beneath this sea: terrestrial Antarctica, so attenuated in life and diversity, is surrounded by a bouquet of species. These life-cloaked plains and rocks, which begin just beyond the reach of the ice and the weather, are mostly beyond my reach as well. Scuba diving becomes dangerous beyond this point. I am observing the frontier outcropping of a strange fauna, which itself is growing perilously close to the lethal instability of the shallows. Until recently, this seafloor was sampled only by traps, dredges, and claws dropped from ships far overhead, and the strange organisms arrived at the surface jumbled and broken and only vaguely hinting of their behaviors. It was rather like studying the trees, foxes, and blue jays of a forest by dropping a bucket from a balloon. But today remotely piloted vehicles bearing television cameras are beginning to scan the deep continental shelf of Antarctica beyond the range of divers, and they are revealing vast plains where every square centimeter is crowded with a wealth of invertebrate species almost as rich as those paradigms of diversity, the tropical coral reefs. Yet coral reefs are lithic factories of photosynthesis, primed with sunlight and surging with energy passing from one organism to the next through frenzied grazing and predation. By contrast, all of the deep Antarctic species are dependent, directly or indirectly, on the dead bodies and organic matter that tumble down the slopes from the algae-cloaked shallows or from the sunlit galaxies of plankton above. Theirs is a borrowed energy. They are the recyclers, a diverse community of necrophages.

*   *   *

The margin of Antarctica is not everywhere a steep edge. Great shelves of ice, 200 to 250 meters thick, overhang thousands of kilometers of coastline. They give birth to the tabular icebergs that characterize the Southern Ocean. The Ross Ice Shelf, roughly the size of France, is the largest, occupying the bottom third of the Ross Sea. It was on the Ross Ice Shelf that Roald Amundsen and Robert Falcon Scott began their treks to the South Pole in 1911. These early explorers called the edge of the shelf "the barrier," because it appeared to be an imposing wall, forty meters high, that ran across the horizon for hundreds of kilometers.[5] The surf pounds the barrier as it would the edge of a continent, and sudden katabatic winds bearing hard snow swirl over its edge onto the sea. The ocean extends for hundreds of miles under the Ross Ice Shelf, and for a curious biologist the abiding question is: what life, if any, can exist in that black, ice-lidded sea? The answer was revealed by one of the most interesting, but least celebrated, explorations of modern times. In 1977 and 1978 a hole was drilled in the Ross Ice Shelf 430 kilometers from the open sea through 420 meters of ice and 237 meters of water, and television cameras, lights, traps, and bottom samplers were dropped onto the seafloor below. The cameras revealed a barren plain that supported an impoverished community of animals, a mere hint of the diverse fauna of the floor of the open sea: a few errant amphipods, copepods, a single *Serolis,* and some shrimp-like crustaceans that were presumed to be mysids and euphausids. There were also a few small antifreeze fish. The sediments below were devoid of life. With no sunlight, no photosynthetic algae, no organic matter from above, these organisms live in a starved world. And because there is no exchange of gases with the atmosphere, the amount of dissolved oxygen is barely enough to sustain life. The water that seeps out from under the Ross Ice Shelf into western McMurdo Sound is almost depleted of oxygen, like an exhaled breath.[6]

Yet, there are pockets of water even more remote, barely accessible even to the imagination, under Antarctica. Radar probes have revealed lakes beneath sectors of the Antarctic ice cap, which in some places is more than two kilometers thick above them. These lakes are below sea level, due to the compression of the continent by its ice shield.[7] I wonder about the contents of those lakes, compressed beneath kilometers of moving ice and solid rock. Are they warmed and kept liquid by the heat of

7

The Worm, the Fish, and the Seal

The invisible worm
That flies in the night,
In the howling storm,

Has found out thy bed
. . . And his dark secret love
Does thy life destroy.

— WILLIAM BLAKE

I CAN VISIT these depths for only thirty minutes before the leaky suit and the cold set me shivering. A hominid whose ancestors evolved in tropical Africa, I have no business here. I inflate my vest and signal Edson to pull me in. Gliding from behind an iceberg, I nearly collide with a Weddell seal. She shows no fear, just curiosity. She stares into my faceplate with big, nacreous eyes; the nictitating membrane that protects the cornea seems almost cataracted. Her vibrissae erect briefly, then flatten once more against her face. Now she swims directly toward me and, with an adroit pulse of a flipper, slides laterally past, taking a long look from her left flank. Finally she takes a breath and holds it, bobbing on the choppy surface, watching. She indolently flicks a nail over her breast. There is no suggestion of aggression. I swim toward her. She backs off, but not far, and watches again.

✳   ✳   ✳

Back at the station, I stand in the shower of generator-heated glacial meltwater for half an hour. Then I eat two plates of black beans and rice, and I am still shivering. But my attitude regarding Antarctica is forever changed: this peninsula is no longer a sere promontory of rock and snow, supporting a few hardy life forms for a few months of the year. I'll make more dives this summer, but still I'll only peek over the edge of the void. I try to imagine the floor of this immense blue bay mantled with life. If only the bay could be made momentarily dry, I could walk across its valley, stepping among its ancient glass-armed sponges and picking up its sea stars.

Instead, like my predecessors before me, I drop buckets from a balloon. In the afternoon after my dive, Claude, Gautier, and I take advantage of the calm and sail a Zodiac through a fibrous slush of ice, first to the shallows on the opposite side of the bay off Cape Hennequin to set the fifteen-meter-long gill net, and then to the center of the bay to drop the trap. Tardin, the Zodiac pilot, is singing "Casey Jones." He once spent three months in Orange, Texas, picking up a ship for the Brazilian navy as well as a passion for American folk songs.

As he steers the inflatable boat between the icebergs, it is easy to become lost in contemplation of them; each berg is unique, never quite to be duplicated. They seem like floating reveries, as vaporous as thoughts, as varied as clouds: dreams incarnate. These musings have seduced every traveler to Antarctica since it was first viewed by humans. One hundred and fifty-six years before me, perhaps also in Admiralty Bay, James Eights wrote,

> It is almost impossible to conceive anything more delicately beautiful than the effect produced by these icebergs, when the sky is free from clouds, and the ocean is at rest; it is then there can be

traced, among the numerous angles and indentations by which they are impressed, all those mingling gradations of color, from the faintest tinge of emerald green to that of the most intense shades of blue.[1]

Drifting off Cape Hennequin, we ballast the net with pieces of rock and buoy it with red plastic floats a yard wide, at just the right equilibrium so that it will hover above the seafloor, fifty-five meters below. The net weighs about a hundred kilograms and must be released slowly so that it doesn't fold onto itself. It slides over the gunwale into the deep blue; only a buoy marks its position on the surface. Now we move on to the center of the bay to drop the trap. The trap is equipped with a transponder that answers when addressed by a sonar blip from the Zodiac; the lag time of the answer tells us the depth. It is buoyed with an air-filled plastic sphere, about the size of a beach ball, which is reinforced so that it won't collapse under the pressure of the water, and is ballasted — just enough to overcome the buoyancy of the sphere — by pieces of iron held in place by an electromagnet. At a signal from the Zodiac, a circuit breaker turns the magnet off, releasing the ballast and allowing the trap to bob back to the surface. The trap has three compartments, with openings four, three, and one and a half centimeters wide. The contents of each will give us clues to the relative sizes of the animals that live on the bottom of the bay. We chum the compartments with pieces of meat and fish and drop the trap into the sea. It is on its own now, and the readings of the light-emitting diode of the depth gauge are slowly rising. Fifteen meters: the maximum depth I reached this morning. One hundred meters: the surface is barely visible now. Three hundred meters — even at high noon on the summer solstice, this is a twilit realm. The trap stops descending at 530 meters. It must be striking the oozing bottom with a puff of sediment. There is no light; the pressure of the water above is fifty atmospheres. The trap is lying far deeper than any human has dived on the Antarctic continental shelf. It is unexplored terrain below us, a void of cognizance onto which we drop our little machine of deceit, offering the gift of a free lunch. Will there be any takers?

✳   ✳   ✳

For two days a tempest lays siege to the bay and we are landbound, unable to retrieve the trap and net. But on the afternoon of the third day, the storm moves on, and we head out to sea as expectant as children on Christmas morning. Using the sonar, we home in on the trap's transponder until, overshooting it, we see the numbers begin to increase. The moment of truth: Claude commands the electromagnet to switch off. Will it release the ballast? There is always the danger of a malfunction; then the trap will remain on the floor of the bay and continue to kill, perhaps for years, its bait steadily renewed by its previous victims. We don't want to harm this bay, only to understand it. To our relief the numbers on the depth gauge begin to decrease, and fifteen minutes later the red buoy bobs at the surface not far from the boat. I imagine the awful journey that the animals inside must have made: from the constant dark and cold, the enormous pressure, to the fickle and sun-blasted surface of the sea. These animals are as adapted to constancy as I am adapted to change. We spill the contents of the trap into pails of water: bright yellow amphipods the size of kernels of corn, a swarm of gray amphipods a centimeter long, several *Serolis*, a few brittle stars, a sea urchin, even a bewildered eight-centimeter-long fish. Two yellow crinoids are clinging to the outside of the trap. Known as feather stars because of the frilled arms they extend into the rain of detritus from above, they must have climbed onto the trap believing that it was rock-steady and permanent. We would find the same types of animals if the trap were set in the shallows. They confirm the peculiar pattern of Antarctica: life on the deep floor of Admiralty Bay is much the same as that in the shallows.

Next, in a strengthening breeze blowing off Stenhouse Glacier, we sail to Cape Hennequin to retrieve the gill net. The bay is becoming choppy. After a long and cold search, we discover that the orange buoy has shifted 400 meters from where we left it. Something powerful moved the net during the three days. Perhaps it was the mysterious deep currents of the bay. Or perhaps a whale; there were minkes in the area this morning.

It takes all of our collective strength to pull the net into the bucking Zodiac. The floorboards are wet and slippery; the net is numbingly cold and soaks our parkas. The net has twenty-six living fish. All but one are marbled notothenids about thirty centimeters long. The other is a Charcot's icefish. Its body, seventy centimeters long, is without scales and is streaming with mucus. The first third of its body is armor-plated head, and through its thin skin I can discern the interlocking plates and sutures of bone. It has a nose like a duck's bill and a mouth with bony lips. It is as pale as a corpse; only its golden-pupiled eye has any color. Most of the fish are helplessly entangled and will quickly freeze and die in the air, which is much colder than the sea. Our task is to extract them and place them in a bucket of water as fast as we can. But in order to manipulate the fine filaments of the net, we have to remove our gloves. It is an excruciating job. Every few seconds I poke my fingers into my mouth to restore sensation and keep them limber. The fish are complacent about being handled and swim slowly in the bucket. This is the imperative of Antarctic marine life: take it slow, conserve your strength.

A bright yellow sea star, seventy centimeters wide, with thirty-nine arms, is clambering up the face of the net. It had been everting its stomach onto the gill-netted fish and slowly digesting them. It is blind, having spent its life feeling its way over the seafloor, eating carrion and extracting clams and other bivalves from the mud. We throw it into the bucket, too.

* * *

The fish are hosts to the same parasites that we observe in the amphipods. Unlike the invertebrates, the fish fauna of the Antarctic is strikingly poor. Of the approximately 20,000 species of fish on Earth, only 120 occur south of the Antarctic Convergence; of these, 95 percent are endemic. Most Antarctic fish, including the marbled notothenias and the icefish, belong to the perch-like suborder Notothenioidei. Surprisingly, only two species, the fifteen-centimeter-long *Pleuragramma* and the two-meter-long *Dissostichus,* are exclusively pelagic and take advantage of the swarms of krill upon which almost everything else in this ocean depends. All the rest are bottom-dwelling species.

The shallow waters around Antarctica once had a fish fauna as diverse as that of any other continental shelf. The 38-million-year-old fossil beds at Seymour Island have revealed catfish, rat-fish, rays, sawfish, and sharks, all temperate or tropical groups that today are absent from or rare in the Antarctic. By that time Antarctica, having sheared away from the rest of Gondwana, was close to its south polar position. Like the terrestrial flora and fauna, most groups of fishes slowly became extinct, except for the new and distinctly Antarctic Notothenioidei, which adapted to the hostile conditions of the coastal waters. As with the surviving terrestrial invertebrates, the challenge for these fishes is to survive subfreezing temperatures without being shattered by ice crystals. Pure water freezes at 0.0 C, but the dissolved salts and minerals of seawater depress its freezing point to as low as $-2.8°$ C.[2] The marine invertebrates on the bottom of the continental shelf commonly encounter temperatures as low as $-2.7°$ C, but since their bodies contain the same concentration of salts and minerals (supplemented by sugars, excretory products, and amino acids) as their external environment does, they don't freeze. Their strategy is simply to stay away from ice, and since ice floats, this usually means staying on the ocean floor. However, the blood and inter-cellular fluids of marine fish are more dilute than the sea, and their tissues freeze at a higher temperature: about $-0.7°$ C. Yet even at winter temperatures, a fish may avoid freezing if it can avoid direct contact with ice. When notothenids are experimentally placed in ice-free seawater, they do not freeze until the temperature drops to $-6.0°$ C. These fish follow the same strategy of supercooling, along with avoiding dirt, ice crystals, and other nucleating particles that precipitate freezing, employed by Antarctica's terrestrial springtails and mites. But these adaptations are not enough. Should an ice crystal even brush against a fish at these temperatures, it can propagate across the skin and penetrate as lethally as any spear.

Obviously, ice is everywhere during the winter, and most Antarctic fish encounter it without ill effects. How do they survive? The question wasn't even asked by physiologists until the 1960s, and the answer was one of the surprising discoveries of Antarctic science: notothenid fish have organic antifreezes in their intercellular tissues and blood. The antifreezes consist of glycopeptides — molecules made of repeating units of sugar and amino acids — that depress the freezing point of water 200 to 300 times more than would be expected from the physical properties of the dissolved substances alone. Notothenid antifreezes operate by a mechanism different from that of the familiar industrial antifreezes or the ethylene glycol in the radiator of a car. The molecules of fish antifreeze adhere to the flat areas of the crystalline lattice of developing ice, forcing any further crystallization to be restricted to curved areas. Because the curved areas have a larger surface area relative to their volume, they shed water molecules to the surrounding sea and freeze at lower temperatures.

Another problem the notothenids faced as the sea gradually cooled over the millennia was that hemoglobin, the iron-bearing molecule that most vertebrates (and some invertebrates, such as the nemertean worms) carry in their blood corpuscles to transport oxygen, becomes inefficient at low temperatures. Hemoglobin is costly, requiring a high investment of energy and nutrients to manufacture. Also it takes a great deal of metabolic energy to push the viscous, corpuscle-laden blood through narrow vessels. To cope with this problem the notothenids gradually became less dependent on hemoglobin and more dependent on the direct dissolution of oxygen in their blood plasma. The most extreme cases of this adaptive anemia are the sixteen species of icefishes, all notothenids in the family Channichthyidae. Icefish have been recognized for decades by Antarctic whalers, who called them "white crocodile fish" because of their large, toothy mouths and pale color; even their gills are creamy white, and their blood is transparent. But it was not until 1950 that biologists discovered that the reason for the pallor was that icefish entirely lacked hemoglobin and had only vestigial numbers of corpuscles. Since then, experiments have showed that only 0.50 to 0.75 percent of the blood of icefish by volume is dissolved oxygen, ostensibly not enough to stoke the fires of a meter-long predator. At first it was believed that the fish compensated for the diminished oxygen-carrying capacity of their blood by minimizing activity and lying in wait for their prey. But now it is known that icefish are as active as any other Antarctic fish and that a few species actually swim briefly into the open sea. Icefish have anatomical compensations for their anemia: a large heart and wide blood vessels, creating a circulatory system of large volume and low pressure. They are also able to build up high temporary oxygen debts in their tissues, to be repaid later during periods of relative inactivity. These compensatory adaptations almost seem to be evolutionary afterthoughts that were jury-rigged to give icefish the energy-saving advantage of thin blood. Now they are relegated to the bottom of the bottom of the world, unable to colonize more benign environments and face competition from their red-blooded brethren. If the tectonic plates bearing Antarctica drifted north again, could these specialists survive the slow warming?

But the notothenids face more immediate threats than the drift of continents. Like krill, they are packets of protein and fat in a starved sea full of hungry neighbors. Weddell seals, leopard seals, elephant seals, toothed whales, petrels, blue-eyed shags, even the red-blooded nemertean worms, hunt them and eat them. It is estimated that seals and birds alone annually consume at least 13 billion kilograms of fish in the Southern Ocean. Unlike the krill, which are only one step removed from the sun's energy, most notothenids are bottom-dwellers that feed on amphipods and other necrophages and are indirectly dependent on the rain of nutrients from above. They have much in common with the bottom-dwelling invertebrates: long lives and the production of a small number of big, yolky eggs that are deposited on the seafloor rather than cast into the hungry plankton. Most notothenids grow slowly, attaining sexual maturity only after four to eight years and reaching maximum size after about twenty years.[3]

✳   ✳   ✳

Back in the cold lab, we place the invertebrates and fish in the aquariums. The amphipods take shelter under the urchins and debris; the thirty-nine-armed sea star plasters itself against the side as flat as can be. The fish are morosely huddled in the corners of the aquariums, awaiting dissection. They harbor the solution to the puzzle of the life cycles of the parasites we are studying. The icefishes in particular are heavily infested by parasites: leeches cling to the skin on the inside of the mouth; copepods and isopods hide beneath the gill chambers; nematodes, cestodes, trematodes, and acanthocephalans invade their guts and organs. Renato will spend the afternoon examining the innards of today's catch. He is skilled in the identification of all these parasites, even their nondescript larvae. Using a microscope, he teases each individual parasite from the tissues — there may be hundreds in a single fish — records it, and preserves it. Some of the parasites are encysted in capsules of connective tissue and have to be placed in caustic reagents to be released. Like pieces of a mosaic, each bears a clue to the interrelationships of the animals in Admiralty Bay.

In a sea where nutrients and energy are scarce, parasites are the ultimate conservatives: they attach themselves to their food supply and withdraw a little nourishment at a time. My special interest is the acanthocephalans, the so-called spiny-headed worms, a few millimeters long, that parasitize seals, fish, and invertebrates. Not truly worms, they are in fact so unrelated to other life forms that they are a whole phylum by themselves. As adults, the acanthocephalans live in the guts of seals; as larvae, they live in the tissues of invertebrates, including several species of amphipods in Admiralty Bay. The larvae cause terrible damage to the amphipods, migrating through their tissues and rupturing organs. If the infestation is heavy, both host and parasites will die. As an ecologist, I am interested in the rate of parasite-induced mortality; it is an environmental force as veritable as predation or food supply. But a good parasite will not kill its host, on which it depends for everything. Virulence is maladaptive. Debilitation, loss of vigor, and decrease in reproductive activity as a result of disease are the more usual results of parasitism. Morbidity is much more difficult to detect than all-or-nothing mortality. How does one measure how many fewer eggs this fish produced because she was infested with worms? How does one know what is "normal" in a sea that is largely unexplored? Obviously, the answers won't come this summer. The best Renato and I can hope for is to add our little sliver of knowledge to the libraries of Antarctica.

But there is beauty in the search. And there is beauty even in parasites, these minuscule animals that accommodate their physiologies to their hosts and manipulate them in extraordinary ways. Take, for example, transmission. The greatest danger in the life of an acanthocephalan (indeed, of any parasite) occurs when it moves from one host to the next. The safest and most economical way to do this is to ensure that the current host is swallowed by the next host; the parasite that can accomplish this feat never has to run the gauntlet of the open sea. And so, toward the end of its residence inside an amphipod (when it comes close to exhausting its host's resources and is therefore imperiling its own life), the larval acanthocephalan induces the amphipod to undergo a remarkable physical and behavioral transformation. The parasite addles the pigment of the amphipod, causing it to lose its cryptic gray coloration and turn brick red. Then, because its immune system is compromised, the amphipod becomes heavily colonized by white yeasts, giving it a mottled appearance. Finally, its instinct to hide beneath rocks and debris is overridden by a parasite-driven imperative to convulsively gyrate through the open water. Utterly conspicuous, it is eaten by a fish in no time.

The fish is simply a paratenic, or transfer, host for the acanthocephalan, which hunkers down in gut or burrows through the intestinal wall and waits — perhaps for years — without developing further. If a seal eventually eats the fish, the life cycle of the acanthocephalan will be completed. But if the fish dies of old age or is eaten by a nemertean (or, indeed, by anything *but* a seal), the acanthocephalan will die waiting. However, the fish is unlikely to live to a ripe old age, and the parasite eventually finds its way into a seal.

Once inside the seal, digestive processes release the acanthocephalan from the fish, and the parasite, now an adult, uses its spiny proboscis to attach to the intestinal wall, where it absorbs nutrients from the passing food. It does not directly parasitize its seal host but instead robs it of food; however, its spines tear tissue and cause serious bleeding. The damage can be severe: hundreds of acanthocephalans have been recovered from a single fish, and thousands from a single seal. The acanthocephalans have no digestive tract; they absorb adundant food directly through their skin from the gut and tissues of their host. Regardless, they are well equipped for reproduction: the males are endowed with a penis, and the females have a complex reproductive tract. Copulation takes place in the seal's gut, and soon after, the female releases hundreds of eggs, which leave the seal's body in the feces. Feces are a nutritious bonanza on the starved ocean floor and are immediately scarfed up by a new generation of hungry amphipods. The life cycle, which may have lasted years, even decades, is complete.

*   *   *

My job this afternoon is to collect fresh excreta from the Weddell seals that wallow on the beaches at Admiralty Bay. Only fresh poop will do; the cold, dry air quickly causes many of the parasites and eggs that I am seeking to encyst, dry up, or jump ship. Rather than waiting for a seal to defecate (which it usually does in the water, anyway), my plan is to startle it, hoping that in its eagerness to get away from me, peristalsis will occur. But how do you scare an animal that has no terrestrial predators and is naturally tame? Weddells, in fact, are so tame that researchers in McMurdo Sound are able to place catheters inside their veins without anesthesia.

Science is often, by necessity, intrusive. I choose a likely candidate, a juvenile female Weddell lying on the beach ahead. Perhaps she is the same seal I encountered in the sea near the iceberg. Full-bellied, she may be ready to relieve herself. Upon hearing my footsteps in the crunchy gravel she raises her head and looks at me, using a single webbed fingertip to scratch herself, then flops her head down and, with a sigh, closes her eyes. She is no more afraid of me in my element than she was in hers. Her fur has water-drop markings that look like the broad brushstrokes of Japanese silk painting. She is basking with her body aligned east to west in order to expose the maximum surface area to the sun's warmth. Where the fur has dried in the sun, its pattern changes from blotched to uniform tawny gray. The fluffed translucent hairs of the dry fur are traps for the sun's energy, converting visible light to infrared radiation and, by internal reflection, retaining it. Each hair is in effect a miniature greenhouse.

Patience is rewarded. I don't have to wait long for the seal to defecate; this growing animal is processing food as rapidly as possible in preparation for winter. Scooping the warm feces into a plastic bag, which I carry inside my coat to keep them from freezing, I take my bounty back to the lab.

✳   ✳   ✳

The seal is one of the young of the season, born on the pack ice of Bransfield Strait. Equipped with protruding incisors, which they use to chew breathing holes in the ice, Weddell seals arrive here early in the spring, following the fissures in the pack. By October most of the females have hauled out on the ice and given birth.

The placentas provide a brief feast for the skuas and giant petrels. Within days the females are in estrus, and the seals copulate, belly to belly, beneath the natal ice floe. The males aggressively defend the ephemeral holes and cracks, which give them access to the females who use the holes. For the first six weeks of her life, this pup lived by her mother's side on the ice, sucking the rich milk and quadrupling her weight, from twenty-five to over a hundred kilograms. By mid-November, when the pack ice was breaking up, the pup was fully weaned, and her mother, already pregnant with the next season's pup, abandoned her.

There are six species of seals in Antarctica and the subantarctic area: the Weddell, Ross, crabeater, leopard, elephant (belonging to the family of "true" seals) and the fur seal (belonging to the family of eared seals — so named because they have external ears — which also includes the sea lions and walrus). Unlike the penguins and the notothenid fish, seals are not evolutionary products of Antarctica's long journey to the south but are relatively recent arrivals. The true seals originated from otterlike ancestors in the North Atlantic; the fur seal is a very different animal, descended from doglike ancestors in the North Pacific. Both the true seals and the eared seals must haul out on either ice or on shore to bear their young, but the true seals are fundamentally aquatic mammals that lack dextrous forelimbs and must undulate over the land like giant grubs. As protection from the cold, they have a thick layer of subcutaneous fat, which makes them even more awkward ashore. The fur seal, by contrast, is able to rear up on its flippers, rotate its hind limbs, and walk — even sprint — on land. It has relatively little blubber but is insulated with a layer of air held in place by dense hair. These adaptations to the cold — the blubber of the elephant seal, which can be rendered into oil, and the luxuriant pelt of the fur seal — have attracted human commerce and so have threatened the number of seals during the past two centuries.

All of these seals except the Ross, which inhabits the dense, closed perennial pack ice, are common in the South Shetland Islands and in Admiralty Bay; like the three species of penguins at

Point Thomas, each seal species makes its living in its own unique manner. As demonstrated by its load of parasites, the Weddell seal eats mainly fish, supplemented by some squids and invertebrates. Since Antarctic fish are mostly bottom-dwellers, Weddell seals must dive deep to exploit them — typically 300 to 400 meters — and they must stay down for over an hour. Admiralty Bay is full of the characteristic underwater vocalizations of hunting Weddell seals, descending Doppler sounds that may travel for kilometers. It is unproven whether the calling seals are echolocating, communicating with other individuals, or both.

The elephant seals probably dive just as deep, but they dive in the open sea and their prey is almost exclusively squid, for which their blunt, peglike teeth are ideally suited. The Ross seal feeds on both squid and fish that it catches beneath the ice. But it is the krill-feeding seals — crabeater, leopard, and fur seals — that are most characteristic of the Antarctic. "Crabeater" is an obvious misnomer, since there are no crabs in the Antarctic. The name was given by the early sealers, who mistook krill for crabs. The crabeater is a filter-feeder. Its lobed teeth, when clenched, make an efficient sieve. Although few observations have been made of feeding crabeaters, it is assumed that they take bites from a swarm of krill, extracting one or two. Crabeaters sometimes feed collectively, which may increase the efficiency of their hunt. Since the decline of the filter-feeding baleen whales, the population of crabeaters, like the populations of krill-feeding penguins, has grown explosively. Until recently it was believed that there could be as many as 30 or 40 million crabeaters, and the population is still increasing; more conservative estimates indicate between 8 and 10 million. Regardless, it is one of the most abundant large mammals — and certainly the most abundant marine mammal — on Earth.

The leopard seal, a close relative of the crabeater, also has filtering lobed teeth. At 500 kilograms, with a large reptilian head and elongated flippers, it is a formidable killer. A far more catholic feeder than the crabeater, it also eats fish, penguins, and, in an ocean without sharks, other seals. Leopard seals patrol the beaches near the penguin rookeries, mugging the adults as they commute from sea to shore, and especially the new-fledged chicks. The ocean that laps the penguin rookeries is as valuable a territory to the leopard seal as the sky above is to a pair of skuas. They also eat young crabeater seals, slicing them with their huge incisors. In Admiralty Bay the bodies of a majority of adult crabeaters show parallel scars, exactly the width of a leopard seal's gape.[4] Mortality of crabeater pups is 30 to 80 percent, a vulnerability that diminishes with the attainment of size and experience after about a year.

The nature of their domain drives the social behavior — and even the morphology — of all the Antarctic seals. Like the Weddell seal, the crabeater, leopard, and Ross seals whelp on the ice. And as with the Weddell seal, the males defend their territory at sea, where speed and agility, rather than size, are advantages. As a result, the females of these species are the same size as the males or slightly larger (at least at the beginning of the breeding season, before lactation depletes their reserves).

The opposite is true among the fur and elephant seals, which pup on the beaches. A large male seal will appropriate a section of shore and, through size, feint, and belligerence, reserve for himself exclusive access to the females that come ashore to calve. This territorial behavior amplifies his reproductive fitness, sometimes extravagantly: a successful male elephant seal, known as a "beachmaster," may defend more than a hundred females (and sire as many pups). A male fur seal may defend thirty or forty females. But since the ratio of males to females among elephant and fur seals is about even (at least at birth), it is obvious that most of the males in these populations are reproductively superfluous.[5] Most of them will leave few, if any, progeny, but those that manage to acquire a stretch of beach will be inordinately successful. This reproductive imperative has led to great selective pressure for large, intimidating size in males. Mature male elephant seals weigh up to 4,000 kilograms and are among the largest of all mammals, but the females weigh only 900 kilo-grams. The ratio is similar in the smaller fur seal: males weigh up to 200 kilograms, females only 50.

Male elephant seals spend most of their terrestrial hours threat-ening, glaring, growling, blubbering, inflating their proboscises, biting, and bashing other males, while the females, gravid with the fetuses of the last season, watch indifferently. I once spent an afternoon on the beach at Stromness, on South Georgia Island, where hundreds of elephant seals haul out every summer amid the rubble of an abandoned Norwegian whaling station. The seals wallowed inside the broken shells of buildings and undu-lated over the cables, platforms, and winches. The beach re-sounded with the guttural declarations of the male seals, sound-ing like long and satisfying belches. The males rose up to batter their opponents with chests and heads. Sometimes two males used each other as supports in order to thrust higher and higher, until they teetered on their bellies like Brobdingnagian sumo wrestlers, bloodshot eye to bloodshot eye, toothy, mouths gap-ing. United, they were taller than I and ninety times my weight, but they were so engrossed in the contest that they ignored me. The battle ended when one collapsed and retreated to the uncon-tested sea.

The victorious beachmaster will never know his offspring, which will be born a year later in early September, very likely on the territory of another male. Uncertain of the paternity of the current season's pups, he takes no interest in their welfare, and in fact may maim or kill the babies by stepping or wallowing on them. The females go into estrus nineteen days after giving birth, and the job of the male is to wait until her behavioral and chemical cues invite him to repeated brief copulations on the hard rock beach. Their union is without delicacy, a study in incongru-ity. Four times her weight, he pulls her close with his flipper and, seemingly enveloping her, strains his ventrum toward hers.

However elaborate and prolonged the male elephant seal's ter-ritorial defense, maternal behavior is simple and abbreviated. The pup, which weighs about forty-five kilograms when born, doubles its weight in eleven days, and quadruples it in twenty-one. By then its mother has entered estrus, and the pup is fully weaned and on its own. Like penguin chicks, elephant seal pups have a surfeit of baby fat and may spend several days or weeks waiting on the beach before entering the sea. Big-eyed and short-nosed, they bear only a slight resemblance to the adults. By the middle of October, just when the phytoplankton are increasing and the sea is full of food, they take the plunge. Relying solely on instinct to hunt for squid, the seals are vulnerable during their first weeks at sea, and many of them die. If squid are abundant in the years to come, the females will first conceive in their fourth year, but in their fifth or sixth year if resources are scarce. Bulls also reach puberty after four to five years, but they don't attain social maturity — the body size and savvy requisite to defending a territory — for at least seven years.

Abbreviated parental care is characteristic of all the true seals in Antarctica. Weddell seal pups are weaned after forty-two days; crabeaters after twenty-eight days. This behavior may be an artifact of their ancestral ranges on the shores of the North Atlantic, where predators roamed the beaches and there was strong selective pressure for a short childhood and a rapid retreat to the safety of the sea.

By contrast, fur seals originated on North Pacific islands where there were few, if any, predators, and fur seal mothers tend to their pups for most of the summer: 110 to 115 days. On South Georgia the pregnant female fur seals haul out in early December, after the elephant seals have mated. By then the bickering male fur seals have delineated their territories, which they monitor — chests inflated, snouts arrogantly pointed skyward, accompanied by falsetto whinnying — from heaps of stones, mounds of tussock grass, even flotsam. The females give birth within two days of making landfall, and eight days later they enter estrus. Once they have conceived, their bodies cannot sustain both lactation and gestation, and the embryos enter a period of delayed implantation and arrested development. The females begin a wearisome cycle of lactation and feeding: four to eight days at the side of the pups, an equal period at sea. As the summer progresses, they spend longer periods at sea, with less frequent nursing bouts. The priority for the mothers is to nourish themselves. Only surplus food is transmuted into milk and, especially during El Niño years (when krill are in short supply), many of the pups perish. It is precisely the prolonged period of infant dependency that has made the fur seal population so vulnerable to hunting by humans over the past two hundred years; when a mother is killed, her baby inevitably starves. The hunter destroys the future of his own industry.

The male elephant and fur seals have now assured their genetic representation in future generations, and they disperse. But it is no surprise that beachmasters suffer enormous stress and injury and live only a few years after attaining social maturity. Obsessed by dominion, a beachmaster disdains food for much of the breeding season, and by the end of the summer he has lost a third of his weight. Malnourished and weak, he may die of wounds, of blindness when an eye is snagged by an enemy's tooth, or of starvation due to broken teeth. But he has passed on his genes — and his self-destructive instinct for territoriality — to dozens of babies.

8

Visions of Ice and Sky

> ... this soul hath been
> Alone on a wide, wide sea:
> So lonely 'twas, that God himself
> Scarce seemed there to be.
> — SAMUEL TAYLOR COLERIDGE

I SLEEP AND DREAM in a place that for most of human history has lain beyond the rim of realization. The witnessing of Antarctica occurred more recently than that of any other continent. Our history texts are steeped in the heroic age of Antarctic exploration, most of which took place early in this century in the sector of the Ross Sea, 2,500 miles west of the South Shetlands. But the first humans to see Antarctica may have been Patagonian Indians or South Pacific islanders who were not part of Western tradition. During pre-Columbian times the whole southern tip of South America was inhabited by a durable group of skin-clad tribes who dived naked in the frigid fjords. They were always accompanied by life-giving fire; even their canoes had hearths. And they may have landed in Admiralty Bay. In 1983 two projectile heads were dredged by Chilean archaeologists from the sediments of sheltered Escurra Inlet, four kilometers northwest of the penguin rookery at Point Thomas. The artifacts were flecked quartz of the type used by the Indians of central Chile until about 1500.

In this time of nationalistic passions about Antarctica, their discovery was greeted with skepticism. How did it happen that they were found on the first dredge attempt? Why were they not encrusted with marine organisms? Was this a ploy by Chile, which annexed the Antarctic Peninsula in opposition to the equally adamant claims of Argentina and Great Britain, to push back the antiquity of its territorial claim to pre-Colombian times? If so, there is a peculiar irony in this, since the Patagonian Indians were driven to near-extinction by the European interlopers who now invoke their priority. Or were the artifacts souvenirs brought to Admiralty Bay, and subsequently lost, by one of the numerous visitors during this century and the last? But it is at least possible that these projectile points were brought to Admiralty Bay by the Indians themselves, in their skin boats. They could have been shoved south across the Drake Passage by an angry storm. The summers in Admiralty Bay would have been bountiful; they could have picked limpets from the cold, shallow sea, hunted seals, and eaten penguin eggs. But without fuel they could not have survived the dark winter. Did they attempt to return? Did they know in which direction Patagonia lay? Could they have compensated for the West Wind Drift and not been luffed eastward into the South Atlantic?

Fables and folklore are, of course, the most enticing historical records, as well as the least verifiable. The Polynesians of the Cook Islands have a legend, passed through the generations by a tribal song, of one Hui-te-rangiora, an adventurer who in A.D. 650 sailed into the far South Pacific, past New Zealand. There he encountered a white island. Hui-te-rangiora had probably never seen ice, so he likened his discovery to the whitest substance that he knew in his tropical realm: arrowroot. He named the sea Tai-uka-a-pia, meaning "the sea with foam like arrowroot." Hui-te-rangiora turned back and lived to tell his tale. He certainly couldn't have reached Antarctica, which lay 2,600 kilometers south of New Zealand, but he may have sailed as far south as Campbell Island. He may have been the first person to see an

Antarctic iceberg. But some Polynesian historians believe that the Cook Islanders hybridized their legends with the tales of missionaries and whalers who visited the region a thousand years after Hui-te-rangiora.

✳   ✳   ✳

Antarctica has haunted the imagination of Western cultures, particularly in the civilizations clustered on the shores of the Mediterranean. Although they had no direct proof of its existence, Ancient Greek cartographers *sensed* Antarctica. And indeed, in terms of its influence on global climate, generating the prevailing winds and ocean currents that pushed commerce across newly discovered seas, Antarctica's vicariate presence was obvious. By 322 B.C. Artistotle had proved that Earth was a sphere by showing that as one traveled north or south the constellations shifted their positions in the vault of heaven. Earth and heaven marched in unison in a matrix of geometry. This same discovery enabled the estimation of latitude and allowed Eratosthenes (circa 250 B.C.) to deduce that the circumference of the Earth was 360 times the distance one had to travel to observe the stars shift one degree. (Sadly, these lessons were forgotten in Christian Europe for over a millennium.) It was therefore obvious to the Greeks that the known "habitable" world occupied about one-quarter of Earth's surface. Did land or sea occupy the rest? The school of Stoic philosophers viewed the other three sectors as necessarily occupied by land masses, endowing Earth with symmetry and balance. One of these continents was Antipodes, the putative Antarctica; starting with the Egyptian geographer Ptolemy (A.D. 150), Antarctica appeared on maps as *Terra Australis Incognita*. Most cartographers considered the south polar continent to be an extension of Africa and depicted it on their maps as a vast white void.

After the European discovery of the New World, occasional drifters claimed to have seen a white land at the bottom of the world, but the confirmation of their discoveries remained as elusive as sea smoke. Almost all the reports came after storms that blew ships far off their courses, as if Antarctica, the spawner of the cyclonic storms that prowled the Southern Hemisphere, was reaching out to be discovered. In 1502 the great Florentine navigator Amerigo Vespucci sailed down the eastern coast of South America from Brazil, which had been briefly explored the year before by Pedro Cabral. Vespucci chronicled his voyages in both maps and letters, and in doing so inadvertently gave the New World his Christian name. One of the reports attributed to Vespucci is a letter written to Piero Soderini, a Florentine bureaucrat, in 1504. In it Vespucci describes sailing southeast into the South Atlantic until "we no longer beheld the stars of Ursa Minor or Ursa Major." Soon "there arose a tempest of so much violence upon the sea that we were compelled to haul down all our sails, and we scudded under bare poles before the great wind, which was south-west, with enormous waves and a very stormy sky: and so fierce was the tempest that all the fleet was in great dread."[1] The storm pushed Vespucci's ship beyond the limits of the known world. "While in this tempest on the 7th day of April, we had sight of a new land, along which we ran for about 20 leagues, and found that it was wholly a rocky coast: and we beheld therein neither any harbour nor any people, because, I believe, of the cold which was so intense" Vespucci's discovery may have been South Georgia, and if so, his was the first European sighting of an Antarctic landmass. However, the authenticity of the letter to Soderini has been disputed by historians, and whether Vespucci and his crew were the first Europeans to see the Antarctic will probably never be known.

In 1503 the French navigator Sieur de Gonneville, a high-born Norman who was attempting to follow the spice-scented track of Vasco da Gama to India, was driven by the usual tempests of the South Atlantic to a new tropical southern land, a place of bounty which he named Southern India. He spent six months there among a people who "asked nothing but to lead a life of contentment, without work." "Southern India" may well have been southern Brazil. On the return voyage, an English privateer captured de Gonneville's ship, yet he managed to return to France with a cargo of skins and decorative feather objects, as well as an Indian "prince" from the newly discovered land. The prince prospered in France, eventually married into the Gonneville family and, as we shall see, 150 years later, one of his descendants was to inspire a whole new generation of Antarctica seekers.

Portuguese and Spanish merchant-explorers forged ever farther south along the eastern flank of South America and soon reached Patagonia. While fleeing the frequent tempests of those waters, Ferdinand Magellan explored every river and bay of what is now Argentinian Patagonia, searching for a sheltered route westward. On October 21, 1520, he reached a rocky promontory, which he curiously named the Cape of Eleven Thousand Virgins, south of which a wide strait headed west from the Atlantic to the Pacific. The strait still bears Magellan's name. He named the wind-swept island south and east of the strait Tierra del Fuego (land of fire), because of the ever-burning fires of the Patagonian Indians that punctuated the dark, wind-swept hills. Magellan proved that the Americas were not attached to *Terra Australis Incognita,* but in doing so he also demonstrated the astounding magnitude of the continents. For fifty years after Magellan, Patagonia was ignored by the Europeans, and its hearty Indians lingered in the Stone Age. The Portuguese concentrated their energies on colonizing Brazil, and the Spanish, having forged a trail across the Isthmus of Panama, found a shorter and safer route to their Andean colonies.

In 1578 Francis Drake, commander of the *Golden Hind* and four smaller vessels, arrived in the Strait of Magellan. Dispatched by the government and merchants of Britain to open commerce (which in his era also meant freebooting) in the South Pacific, he was ordered to explore the coasts of all southern continents that were "not in the possession of any Christian Prince." On August 17 the *Golden Hind* reached the Cape of Eleven Thousand Virgins, and Drake headed west through the strait. Seven days later, while bivouacking on an island in the middle of the strait, the explorers observed, in the words of the Reverend Francis Fletcher, chronicler of the voyage, a "great store of foule which could not flie, of the bignesse of geese, whereof we killed in lesse than one day 3,000 and victualled ourselves thoroughly therewith."[2] These were Magellanic penguins, which still are found abundantly in Patagonia, and they were the first penguins described by Englishmen.[3] By September 6, the fleet had safely entered a bay on the Pacific side, but the next day a storm separated the *Golden Hind* from the rest of the fleet (Fletcher melodramatically called the site "The Bay of Severing of Friends") and pushed her southward along the Fuegian coast. They traded with Indians, "both men and women in their Canoas, naked, and ranging from one island to another to seeke their meat."[4] Eventually the *Golden Hind* reached the open sea again. Fletcher wrote, "At length we fell with the uttermost part of land towards the South Pole . . . without which there is no maine nor Iland to be seene to the Southwards, but the Atlanticke Ocean and the South Sea [Pacific Ocean], meete in a most large and free scope." This was Cape Horn, the absolute tip of South America, and the 900-kilometer-wide strait to its south has since been known as the Drake Passage, still regarded as the most tempestuous body of water on Earth. The *Golden Hind* drifted beyond 57° S. Seeing no evidence of land, let alone objects of commerce or pillage, in these high latitudes, Drake concluded that it would be more profitable to swashbuckle up the South and North American coasts, raiding the Spanish colonies, which in turn were stealing booty from the Indians. But because of Drake's voyage to nowhere, Fletcher had the conceit to declare that the expedition had "altered the name [of Antarctica] from *Terra Incognita* to *Terra nunc bene Cognita.*"

The long decline of the Fuegian Indians, due to disease and the usurpation of their land, then began. In 1582 the Spanish, realizing the vulnerability of their Pacific colonies to attack through the Strait of Magellan, sent a garrison of 3,500 men and women to

Patagonia. By the time the garrison reached its goal, only 400 men and 30 women survived, the rest having succumbed to English privateers and storms at sea. Landfall didn't improve their fortune; by 1592 only 23 cold and starving pilgrims were left. While attempting to return to Spain to organize a rescue mission, their leader was kidnapped by none other than Sir Walter Raleigh. Ironically, the person who eventually rescued the survivors was also an English freebooter, Thomas Cavendish, captain of the *Galeon*. In August of 1592 another of Cavendish's fleet, the *Desire*, a merchant vessel under the command of John Davis, was blown by a winter storm eastward from the coast of Patagonia to discover the Falkland Islands.

Antarctica waited undiscovered and would remain so for 227 more years. But there were some near misses; if the currents had shifted slightly or the cyclonic storms lingered a little longer in the Drake Passage, discovery might have taken place during the first centuries of European exploration of the New World. During a tempest in 1603, the Dutch ship *Blijde Bootschap* ("Blithe Tidings"), under the command of Dirck Gherritz, a buccaneer-trader en route to the Indies, was blown south from Cape Horn to 64° S, from where Gherritz spied "a high land with mountains covered with snow, resembling the land of Norway" that extended all of the way to the Solomon Islands.[5] Obviously Gherritz didn't sail as far west as the Solomons, and no enduring proof of the adventure has survived. But for over a century the putative southern continent was known as "Gherritz Land," and some historians have suggested that Gherritz had in fact discovered the South Shetland Islands. In any event, until it is substantiated, the tale will be relegated to the apocrypha of history along with the song of Hui-te-rangiora.

Two generations after Gherritz, the tale of Sieur de Gonneville, overshadowed in the excitement of a period of dramatic discovery, was resurrected. In 1663 the great-grandson of the Indian prince who sailed with de Gonneville to France published a celebrated chronicle of his ancestor's voyage, much embellished by the passage of time and borrowings from the accounts of other explorers. The image of "Gonneville Land," cloaked with tropical vegetation and peopled with innocent natives, soon became synonymous with *Terra Australis Incognita,* and this primitivist idyll spawned new romantic interest in the southern continent.

In April 1675, Anthony de la Roché, a London merchant of French descent, rounded Cape Horn on a voyage from Peru to Europe. De la Roché was in command of two ships, with a combined crew of fifty-six men. The usual tempest, a northwesterly gale, struck them when they were rounding the Horn, and the ships were driven off course into the South Atlantic. Winter was setting in, and the crew became apprehensive, but soon they encountered an island with snow-capped mountains and a sheltered bay. This was probably the southern tip of South Georgia. De la Roché hove to in the bay for fourteen days, while the tempest battered the open sea. Finally, in clear weather, de la Roché continued to the southeast and encountered a group of small islands, probably the Willis Islands or the Clerke Rocks. De la Roché assumed these to be the tip of the southern continent, a discovery that was in fact to elude explorers for another 145 years. Regardless, if one discounts the authenticity of Vespucci's letter and of Gherritz's account, it was the first European sighting of an Antarctic landmass, albeit not the continent itself.

Soon after, from 1683 to 1685, the *Batchelor's Delight,* under the command of Abraham Cowley, a good-natured English merchant and adventurer, sailed around the world. While rounding Cape Horn, the crew were "discoursing of the intrigues of women," when a violent storm blew them 60° 30′ S, practically to the South Shetland Islands. Cowley wrote that the sea was "so extreme cold that we could bear drinking three quarts of Brandy in twenty-four hours each man, and nor be at all the worse for it." He concluded that "discoursing of women at sea was very unlucky and caused the storm."[6]

De Gonneville's fancy incited outright fraud when, in 1772, Yves-Joseph de Kerguélen-Trémarec, a Breton nobleman, sailed from Mauritius with two small ships, the *Fortune* and the *Gros*

Ventre, under orders of the court of Louis XV to search for "a very large continent to the south." On January 6, 1773, they happened on several ice-locked islands, shrouded in a frozen fog, in the southern Indian Ocean. Kerguélen, aboard the *Fortune*, never set foot on his discovery (which, regardless, was eventually named after him), choosing instead to lie timidly offshore.[7] Eventually he dispatched a longboat to one of the island's few beaches. A storm, accompanied by a dense fog, set in, and the longboat collided with and practically sank the *Gros Ventre*. Kerguélen, abandoning both the longboat and the *Gros Ventre*, returned to Mauritius and then to France with claims of having discovered "La France Australe." He wrote:

> The lands which I have had the happiness to discover appear to form the central mass of the Antarctic continent. . . . The latitude in which it lies promises all the crops of the Mother Country from which the islands are too remote to derive fresh supplies. . . . No doubt wood, minerals, diamonds, rubies, precious stones and marble will be found. . . . If men of a different species are not discovered at least there will be people in a state of nature living in their primitive manner, ignorant alike of offense or remorse, knowing nothing of the artifices of civilized society. In short La France Australe will furnish marvelous physical and moral spectacles.[8]

Obviously, Kerguelen was a masterful practitioner of those very artifices that he condemned.

In the meantime, François Alenso Comte de St. Allouarn, the dutiful captain of the tiny *Gros Ventre*, having lost sight of the *Fortune*, made a successful landing on the island, claiming it in the name of Louis XV. Ever faithful to his commission, he continued across the Indian Ocean in search of the southern continent. Instead he encountered the western coast of Australia and proceeded north to Timor and Java before returning to Mauritius.

Louis XV, the penultimate French monarch before the revolution, eagerly embraced Kerguélen's claims, and before word of the nobleman's fraud reached France, quickly outfitted an expedition of 700 sailors and colonists aboard a fleet of three ships, the *Roland,* the *Oiseau,* and the *Dauphine,* to La France Australe, under Kerguélen's command. The crew included Kerguélen's mistress, Marie-Louis Seguin, a prostitute from the infamous Rue de Siam in Brest, who probably introduced the oldest profession to the Antarctic regions. Kerguélen's second voyage was as inept as his first. An epidemic of fever soon broke out on *Roland,* in whose hold was discovered "a prodigious quantity of white worms."[9] The ships hove to in the vicinity of the islands for thirty-three days and, as before, Kerguélen never set foot on his new land. He delegated that duty to Charles de Rosnevet, the captain of the *Oiseau.* Rosnevet landed at a place he named Sea-Lion Bay and deposited two sealed bottles containing parchment declarations of the French claim to the islands. On his return to France, Kerguélen, his illusion of La France Australe evaporated, wrote: "It seems quite clear that this region is as barren as Iceland and even more uninhabitable."[10]

But Kerguélen's contrition was not enough for the French court. His folly proved acutely embarrassing to the French navy, and he was brought before a naval court-martial and accused of, among other things, conduct unbecoming an officer, abandoning both the longboat and the *Gros Ventre,* permitting pestilential conditions, and smuggling the prostitute aboard his ship "for his pleasure." In 1774 he was condemned to twenty years at the gentlemen's prison at Saumur. When revolution swept France four years later, Kerguélen, by then regarded as a victim of the monarchy, was released. His rank restored, he spent the rest of his days as harbor manager in Brest (presumably close to the pretty Marie-Louis).

✳   ✳   ✳

And so *Terra Australis nunc bene Cognita* remained unproven, its shores littered with a wreckage of myth and false claims. But in the late 1700s an era of determined scientific exploration of the Southern Ocean began, initiated by James Cook, who was perhaps the most accomplished seafaring explorer of all time.

Cook was the son of a humble, landlubbing Yorkshire farmhand, who began his career as a sailor, working his way up to captain, aboard the sturdy colliers that plied Britain's stormy coastal waters. He relinquished that pedestrian career to start anew in the Royal Navy, again working his way through the ranks to captain. By 1772, already famous for having led an expedition that mapped eastern Australia and New Zealand, Captain Cook set sail to discover the postulated southern continent. "Whether the unexplored part of the *Southern Hemisphere* be only an immense mass of water," wrote Cook, "or contain another continent, as speculative geography seemed to suggest, was a question which engaged the attention, not only of learned men, but most of the maritime powers of Europe." Cook decided to circumnavigate the Earth at approximately 60° S and "to put an end to all diversity of opinion about a matter so curious and important." Imperial England, soon to lose its American colonies, was anxiously seeking more dominion, and Cook's secret orders from the Admiralty read,

> You are to observe the Genius, Temper, Disposition and Number of the Natives or Inhabitants, if there be any. . . . You are with the consent of the Natives, to take possession of convenient Situations in the Country in the Name of the King of Great Britain, and to distribute among the Inhabitants some of the Medals with which you have been furnished to remain as Traces of your having been there.[11]

Cook was paid six shillings a day to lead this ambitious expedition. The Admiralty provided him with two ships, both Yorkshire colliers, the *Resolution,* with 110 men and Cook as captain, and the *Adventure,* with 80 men and Tobias Furneaux as captain.[12]

The expedition's scientific agenda was prepared by the Astronomer Royal and the Royal Society, which sent two astronomers, William Wales and William Bayly, both very competent observers, on the expedition. One of their priorities was to measure longitude accurately. Although latitude had been estimated since the time of the ancient Greeks by the declination of the stars on the horizon, the measurement of longitude required measuring time by chronometers that were accurate to within seconds. The issue was so important in England, now an expanding maritime power, that in 1714 Parliament had established the Board of Longitude, a commission to foster experiments in measuring longitude at sea. A prize of £20,000 was offered to the first person to measure longitude to within forty-eight kilometers after a voyage of six weeks. In 1762 this small fortune was won by John Harrison, a master English watchmaker. The *Resolution* was equipped with a copy of the award-winning timepiece designed by Harrison, which functioned flawlessly throughout the voyage. Wales took daily measurements of latitude and longitude. He also made various astronomical observations; measured wind velocity and direction, air pressure, variations of the compass, and the temperature of the air and the sea at different depths; and described new lands, their harbors, and tides.

The expedition was the first systematic scientific exploration of the Southern Ocean, in its time equivalent in scope and novelty to the flyby of the outer planets of the solar system by the Voyager spacecraft. Cook's ships departed from Cape Town on November 23, 1772, and sailed eastward across the southern Indian Ocean, farther south than any previous voyagers (71° 10′ S). They were searching for land but found none, proving that Kerguélen's France Australe could not have been a northern outlier of the southern continent. On January 17, 1773, the *Resolution* and the *Adventure* became the first vessels to cross the Antarctic Circle.

Twenty-two days later the ships were separated in a storm, and Cook, on the *Resolution,* proceeded alone. On February 23, somewhere near what is now named the Davis Sea in the Indian Ocean sector of Antarctica, in what is still one of the most remote and least visited parts of the planet, Cook wrote,

We now tacked, and spent the night, which was exceptionally stormy, thick and hazy, with sleet and snow, in making short boards. Surrounded on every side with danger, it was natural for us to wish for day-light. This, when it came, served only to increase our apprehensions, by exhibiting to our view, those strange mountains of ice, which in the night, we had passed without seeing.[13]

In the meantime the *Adventure* explored the southern and eastern coasts of Tasmania and then, by prior agreement, continued to Queen Charlotte Sound, New Zealand, where she was reunited with the *Resolution*. During the austral winter the ships journeyed to Tahiti and to the Tonga Islands, both noted for their friendly inhabitants. Once again the ships were separated by a storm, and the *Adventure* sailed back to New Zealand. This time the local Maori tribesmen were not so hospitable: they killed and ate ten of her crew. The following summer a demoralized Furneaux hied across the Pacific, rounded Cape Horn, and continued on to England. During January 1774 the *Adventure* passed only forty-five miles northeast of the South Shetland islands, probably just off the northern coast of King George Island. Had her course been twenty miles farther south, her watch probably would have sighted the glaciated interior of the island.

Come spring of 1774, Cook once more sailed alone to the high latitudes of the South Pacific, to the edge of pack ice. The *Resolution* crossed the Antarctic Circle two more times. But in spite of his perseverance, Cook could not push the Yorkshire collier any deeper into the pack ice. On December 15, at the northern fringe of the Ross Sea, Cook wrote in his journal:

According to the old proverb a miss is as good as a mile, but our situation requires more misses than we can expect, this together with the improbability of meeting with land to the South and the impossibility of exploring it for the ice if we did find any, determined me to haul to the north. . . . We frequently, notwithstanding all our care, ran against some of the large pieces, the shoks which the Ship received thereby were very considerable, such as no Ship could bear long unless properly prepared for the purpose.[14]

By December 23, the *Resolution* was in

> a strong gale which froze to the Rigging as it fell, making the ropes like wires, and the sails like boards or plates of metal. The sheaves also were frozen so fast in the blocks, that it required our utmost effort to get a top-sail down and up; the cold so intense as hardly to be endured. . . . Under all these unfavourable circumstances, it was natural for me to think of returning more to the north; seeing no probability of finding any land here, nor a possibility of getting farther South.[15]

The *Resolution* turned north once again, to Tahiti, the Tonga Islands, and beyond, to the European discovery of New Caledonia and the New Hebrides. Ironically, Cook, a man of iron constitution and determination who was indefatigable in the Antarctic, became extremely ill in the relatively salubrious tropical Pacific and was laid up for a month with a malady that included, among other symptoms, a day and night of violent hiccuping.

In spring 1775, after another winter in New Zealand, Cook crossed the Pacific to Cape Horn and continued east. Within days the *Resolution* encountered South Georgia Island. He explored the eastern coast of the island, charting the bays that would 135 years later become the sites of the great whaling stations. Believing, like de la Roché, that the island must be a northern extension of the southern continent, Cook sailed southeast until he rounded the southern tip of the island. When he realized that South Georgia was not an appendage of Antarctica but an island, he named the place Cape Disappointment. Although winter was impending, the resolute Cook continued southeast. In a few days he discovered more ice-locked islands, which he named the South Sandwich Islands in honor of the Earl of Sandwich, one of his sponsors. These also proved finite and disappointing, and after a week he turned north. By the time she reached England, after a voyage of three years and eight days, the *Resolution* had traveled 97,000 kilometers, but Cook had failed even to gaze on the Antarctic continent. Why was this consummate explorer denied that discovery?

Cook's journal gives us a clue. He described the Antarctic as having an "inexpressibly horrid aspect . . . a country doomed by nature never once to feel the warmth of the sun's rays, but to live buried in everlasting snow and ice." He couldn't have known (indeed it was beyond the ken of science until the 1960s) that at the time of his voyage the Earth was near the end of the Little Ice Age, a period of record low temperatures that began in the mid-1600s. In Europe these were centuries of advancing mountain glaciers, frequent severe winters with low snowfall, cold springs, and short summers that wrought havoc on crops.[16] The effect of this long chill on Antarctica is not known, but it can be assumed that the pack ice extended farther out to sea than it does today. Cook's itinerary aboard the *Resolution* was generally south of 60°, in the daisy chain of cyclonic storms that ply the Southern Ocean. By contrast, Furneaux and the *Adventure,* during her two separations from the *Resolution,* tended to stay a few degrees farther north, above the track of the storms. So Cook was hounded by storms and pushed back by insistent pack ice during those three exceptional summers. His meteorological records for the summer of 1772–73 describe only 3 percent of the days as "fine" or "clear," and during the summer of 1773–74 it was only 1 percent.

Cook did not discover Antarctica, as he had set out to do, but by finding nothing save a few ice-locked islands in the high southern latitudes, he may have made his greatest discovery: the circumpolarity of the Southern Ocean. Cook showed that the far-flung fragments of land—La France Australe, South Georgia, the South Sandwich Islands — were not part of the Antarctic continent and, if the continent did exist, that it was not a temperate, peopled place but ice-locked and remote from the rest of the world. He lamented, "The risque one runs in exploring a coast, in these unknown and icy seas, is so very great, that I can be bold enough to say that no man will ever venture farther than I have done; and the lands which may lie to the South will never be explored." Defeated by the long-term cycling of global climate, itself inscrutably controlled by the orientation of Earth toward the sun — concepts that would not be understood for two hundred more years — Cook complained, with a lack of vision uncharacteristic of such a consummate explorer, "If any one should have resolution and perseverance enough to clear up this point, by proceeding farther to the south than I have done, I shall not envy him the honour of discovery, but I will be so bold to say that the world will not be benefited by it."

* * *

Suddenly and accidentally, at dawn on February 19, 1819, the existence of Antarctica was at last proved. Captain William Smith, owner and master of the brigantine *Williams,* a merchant vessel from Blythe, England, had been typically beset by a storm at Cape Horn while trying to ply his way from Montevideo, Uruguay, to Valparaíso, Chile. Like Cook's ships, the *Williams* was a collier from northern England, and although only 216 tons, she was a snow brig, specially equipped with additional topsails that enabled her to adapt to a variety of winds and seas. Although longitude had been accurately estimated for nearly fifty years, Smith was one of the few navigators of his time to use a chronometer to calculate it. Perhaps bolstered by the extra confidence that his ship and instrument gave him, he sailed south into the swelling seas of the Drake Passage to avoid the treacherous islands at the tip of Patagonia. On the night of February 19 he crossed the path Furneaux had taken in the *Adventure,* forty years before, and at seven A.M. he viewed "land or ice . . . bearing South-east by south [at a] distance [of] two or three leagues — strong gales from the South West accompanied with Snow or Sleet." Soon the weather moderated, and by four P.M., at latitude 62° 01′ S, Smith discerned "land bearing from S.S.E. to S.E. by E. distance about 10 miles, hove to, and having satisfied ourselves of land hauled to the Westward and made sail on our voyage to Valparaiso."[17] With this desultory observation, the existence of the South Shetland Islands was recorded by somebody who could document the discovery.

Upon his return to Valparaíso, Smith made a formal report of his discovery to Captain W. H. Shirreff, the senior British naval officer in the port, but according to one of Smith's crew, J. Miers, Smith was "ridiculed . . . for his fanciful credulity" and for resurrecting the idea of the mythical southern continent.[18] The Spanish colonies in South America were in revolt, and the British authorities, occupied with maintaining their influence in a changing political climate, had few resources to spare for exploration. But Smith, described by a contemporary as "very upright . . . although inclined to be a little pompous," and who wore "his own black hair," instead of a wig, persevered.[19] While returning to Montevideo the following June, he again sailed south from Cape Horn. However, it was the height of winter and the *Williams* was beaten back by the pack ice long before nearing Antarctica. The crew saw plenty of ice but no obvious land. By the time the *Williams* arrived in Montevideo, some of Smith's crew had concluded that they had sighted only a huge iceberg the previous February. Not surprisingly, the British authorities in Montevideo were no more interested than Shirreff had been.

However, the American sealing captains in Montevideo and nearby Buenos Aires immediately recognized the importance of Smith's discovery. These businessman-adventurers were the keystone of a commerce in seal pelts that involved three continents. Sealing had begun in the United States soon after the Revolutionary War, when one Lady Haley of Boston invested in the *Neptune,* a huge ship of about 1,000 tons, for a voyage to the Falkland Islands to hunt fur seals and sea lions (the latter known in the vernacular of the time as "hair seals"). The crew of the *Neptune* killed about 13,000 seals, the salted skins of which were sold in New York for fifty cents each. From New York the sealskins were shipped to Asia and misrepresented as "sea otter" skins.

The hunting of sea otters, prized for their soft pelts, was just starting in the Pacific Northwest. But fur seal pelts were armed with stiff guard hairs that made them scratchy and uncomfortable. The Chinese had invented a method of removing the guard hairs, and in Canton the partially depilated sealskins were selling for about five dollars each. Soon a triangular trade developed between the United States, South America, and China. The American traders returned from Canton with tea, silk, and ceramics; commercial dynasties on both sides of the Pacific were built on this trade. On one voyage the *Neptune* returned to New York with $260,000 worth of Chinese goods, a vast fortune for the time.

The seals of Patagonia, the eastern Pacific, the Falkland Islands, and South Georgia were doomed. From 1793 to 1807 approximately 3,500,000 fur seal skins were taken from just one small island, Más Afuera (now known as Isla San Ambrosio), off the Pacific coast of Chile. Since the War of 1812 Americans had hunted fur seals and sea lions on South Georgia, but by Smith's time there were few left. By necessity, the sealers turned to the less profitable elephant seals, which were still abundant on South Georgia. Elephant seals did not have valuable pelts, but their blubber could be rendered into an oil which, although inferior to sperm whale oil as a fuel and lubricant, found a ready market in newly industrialized Europe.[20]

By the early 1820s, the supply of fur seals had dwindled so much that pelts were selling for seven or eight dollars each in China. The American captains were seeking new sealing grounds to the south, and in particular the mythical Aurora Islands, which were reputed to have numerous seals and which, like Antarctica itself, always seemed just beyond the grasp of discovery.

The American sealing captains in Montevideo offered to charter the *Williams* if Smith would guide them to the islands he had discovered. But Smith, a patriot, refused. He wrote, "Your Memorialist having the Good of his country at heart (if any should be derived from such Discovery) and as he had not taken possession of the land in the name of his Sovereign Lord the King resisted all the offers from the said Americans, determined again to revisit the new-discovered land."[21]

In September 1819, the *Williams* sailed south a third time, and on October 14, land was again sighted. Smith named the islands the New South Shetlands, and explored their coasts for three days. On October 17 his first mate and a small crew landed a tender at the northeast tip of what is now King George Island. Smith named the promontory the North Foreland. His crew planted a board with the Union Jack in the frozen beach, buried a bottle of coins of the realm, and with three cheers, "your memorialist landed and took formal possession of the new discovered land."²² And then the peculiar polar hallucinations set in once more. Smith reported, just as had Gherritz and Kerguélen, that he could "perceive some trees on the land to the S. W. of the Cape." Was this wishful thinking, or did the play of fog, glacier, and light deceive him? Crewman Miers also reported seeing "ornithorhynchus" (platypus) and "sea otters," but these can be easily explained as misidentified penguins and seals.

This time the authorities in Valparaíso were compelled to believe Smith. Captain Shirreff himself chartered the *Williams,* appointing Edward Bransfield as master and Smith as pilot. The orders from Shirreff to Bransfield were to report the "uncommon abundance of Sperm Whales, Otters, Seals, etc., upon the Coast and the Harbours . . . [and] if [the islands] should already be inhabited, the character, habits, dress, customs and state of civilisation of the inhabitants."²³ Obviously, Shirreff had not read Cook's journal and instead naively believed Smith's tales of a temperate Antarctica. Smith and Bransfield sailed from Valparaiso for the Antarctic on December 20, 1819, and soon entered the Drake Passage, where, in the words of Adam Young, the medical officer aboard the *Williams,* they were "almost constantly harassed with baffling winds and calms."²⁴

That same summer British and American sealing ships were headed south as well. In the gossipy emporiums of Spanish America, the word of the tantalizing southern islands had spread like a virus. Besides Shirreff, Smith had told a few of his colleagues, all British subjects, of course, about his discoveries. The Americans had no way of knowing exactly where the newly claimed British sealing grounds were. Many assumed that the new lands were the illusory Aurora Islands. But a chance encounter in the Falklands provided the Americans with the information necessary to poach in the new territories. In January 1819 the American sealer *Hersilia,* under the command of Captain Sheffield of Stonington, Connecticut, put into the Falklands. Sheffield's second mate was Nathaniel Palmer. Only twenty years old, "Young Nat" was already a veteran of seven years at sea, having first served as a sailor on a blockade runner during the War of 1812. He was left in the Falkland Islands to hunt feral cattle and to obtain other provisions while the *Hersilia* made a short and unsuccessful excursion in search of the Aurora Islands.[25]

While Palmer was waiting for the *Hersilia* to return, the British brig *Espirito Santo,* on its way to the South Shetland Islands, also put into the Falklands. Palmer, who knew the harbor, was asked to pilot her to a safe anchorage. The captain of the *Espirito Santo,* who had probably heard of the South Shetlands from Smith, told Palmer that he was headed to a place where there were thousands of seals, but he refused to divulge the exact location. The amiable young Palmer somehow managed to guess the other ship's destination, and three days later, when the *Hersilia* returned, Palmer convinced his captain to hastily outfit her for a sealing voyage to the unknown southern waters. It was only a four-day sail to the South Shetlands, and on January 18 the *Hersilia* was in sight of what is now named Smith Island. On the twenty-first, the crew of the *Hersilia* was sealing alongside the startled crew of the *Espirito Santo* on Rugged Island, a small key off what is now called Livingston Island. There were abundant seals for all. That summer the crew of the *Hersilia* killed 8,868 fur seals and would have taken more had they not run out of salt. Leaving the salted skins to dry in South America for eventual shipment to China, they returned to Connecticut to spread the news.

In the meantime, 300 miles to the west, the *Williams* was plying south from Valparaíso. On January 17 she arrived at Start Point, a promontory of Livingston Island. There was no sign of the American sealers, who would arrive three days later. From offshore, the island's "limits could scarcely be distinguished from the

white clouds which floated on the tops of the mountains," wrote
Adam Young.[26] But the coast was so fouled with submerged rocks
that Smith and Bransfield did not attempt a landing and instead
sailed northeast to King George Island; skirting to the leeward,
southern side of the island, they entered what is now King George
Bay. "Words can scarcely be found to describe its barrenness and
sterility," wrote Young.

> Only one small spot of land was discovered on which a landing
> could be effected on the Main, every other part of the bay being
> bounded by the same inaccessible cliffs. . . . Nothing was to be
> seen but the rugged surface of barren rocks, upon which myriads
> of sea-fowls had laid their eggs. . . . These birds [penguins] were
> so little accustomed to the sight of any other animal, that, so far
> from being intimidated by our approach, they even disputed our
> landing, and we were obliged forcibly to open passage for our-
> selves through them.[27]

After breakfast on January 20, 1820, Bransfield planted the
Union Jack for the second time on King George Island, "in the
name and behalf of H. M. George IV his heirs and successors."
Williams hoisted her ensign and pendant and fired her guns. But
the inhospitality of the place soon became apparent. Thomas
Bone, a midshipman on the voyage, remarked that the ceremony
was invaded by "the most intolerable stench I ever smelt," originating from the large penguin rookery upwind from the bay.

Smith and Bransfield then sailed southwest, past what are now
named Nelson, Robert, Greenwich, and Livingston islands, to
Deception Island. They surveyed the coasts of Deception and of
Trinity Island. On a clear day, the Davis Coast of the Antarctic
Peninsula is clearly visible from Trinity Island. Smith and Bransfield did not land on the peninsula, and even if they glimpsed it,
they had no way of understanding the implications of the discovery. It was not until 1937 that geographers knew the peninsula
was connected to the Antarctic mainland.

Young recorded "very great numbers of whales with which we

were daily surrounded; and the multitudes of the finest fur-seals and sea-lions which we met both at sea and on every point of the coast, or adjacent rocky islands, on which we were able to land."[28] According to Young, there were so many fur seals at North Foreland alone that a captain could retire after just one voyage.

Upon returning to Valparaíso and being released from his contract to the British crown, Smith outfitted the *Williams* as a sealer, with plans to return in the summer of 1820–21 and cash in on the bonanza. But the word of seals in the new lands, so assiduously sought, was on every sealing captain's tongue. Historian H. R. Mill, looking back at those years in 1905, wrote: "The southern summer of 1820–21 was a dark one for the fur-seals whose ancestors had basked upon the shores of the South Shetlands for untold centuries, following their quaint semi-civilized life and pursuing their patriarchal customs of war and love undisturbed by any being capable of contending with them."[29] The free-for-all commenced the following summer. Antarctica, at last perceived by humans, was facing catastrophe.

9

The Indifferent Eye of God

I made a jotting of the men flinching [flensing] him; as a piece of color the effect was gorgeous — masses of scarlet, dazzling white and blue sea. The snuffling of the seal, and the sound of the blood spouting and fizzling into the snow, with the crisp sound of the steel in the quivering flesh was hardly nice, and when the red carcase set up and looked at itself, I looked up to see if God's eye was looking.

— W. G. BURN MURDOCH

A CROSS ADMIRALTY BAY from the Brazilian station is Cape Hennequin, an ice-free shoulder of beach and steep brown hills. At the tip of the cape the entire vertebral column of a whale crosses the pebble beach like a picket fence. At the southern end of the beach, the basalt lava has cooled into polygonal columns of black rock, looking like the ruins of an ancient civilization. The rock faces are spattered with orange and yellow crustose lichens. Cape Hennequin, the site of a petrified southern beech forest, 16 million years old, bears subtle clues to the history of Admiralty Bay. Every spring new mudstones, bearing the impressions of fossil leaves, wash down the hillside like a stone flurry of autumn leaves. It is also the site of a much more recent ruin, the remains of a sealers' camp. You have to look hard to see it: a tumbled stone wall beneath the cliff face, a few whale ribs that may have been used as ceiling beams.

During the summer of 1820–21 there were at least forty-nine sealing vessels in the South Shetland Islands. Most were British and American, but a few came from as far away as India. At least five of the ships hunted on the beaches of King George Island. Poor William Smith, returning on the *Williams* to the islands he discovered, was appalled at the jam of ships. "To your Memorialist's surprise," he wrote, "there arrived from 15 to 20 British Ships with about 30 sail of Americans." Smith estimated that there were about 1,200 British, and as many American, sealers working in the islands that summer.

This mob of sailors left ruins throughout the South Shetlands. One of their favorite sealing beaches was Byers Peninsula, a large ice-free area at the western tip of Livingston Island.[1] Three kilometers offshore is Rugged Island, and across Barclay Bay, sixteen kilometers to the northeast, are Cape Shirreff and Desolation Island, all important ice-free areas that were densely populated by fur and elephant seals. Today the ruins of sealers' refuges still survive. Scattered like the tossed bones of a fortune teller, they divine the details, daily chores, and routines of the first men in the Antarctic. Their structures were rude and temporary, often made of the flotsam found on the beach. Most of the shelters had three walls constructed of boulders and caulked against the insistent wind with scraps of sealskin; the fourth wall was usually a rock face. Whale ribs, spiked with rusting nails, were used as beams to support a canvas sailcloth roof. Whale vertebrae were used as chopping blocks and furniture, and even today are arranged as tables and seats, just as the sealers abandoned them 150 years ago. Seal bones, penguin bones, and Antarctic limpet shells litter some of the sites, indicating that the sealers lived off the land. There are the sundry items of daily life: the mandible of a pig (which suggests that the sealers ate salt pork), a stone hearth, a blubber lamp, a few barrel staves and hoops, iron spear heads, leather boots with leather laces, canvas and sealskin moccasins, the sleeve of a woolen jacket.

The sealers were secretive about the location of their sealing beaches and the means of getting there, and for this reason they kept only minimal journals and logs. In 1909 Edwin Swift Balch, an American historian who recorded the American exploration of the South Shetland Islands for patriotic reasons, was privy to some letters and papers belonging to Mrs. Richard Fanning Loper, Nathaniel Palmer's niece. Fifty-nine years before, the Palmer family house in Stonington had been destroyed by fire and most of the family's Antarctic records lost, but one of the surviving documents was a manifest for the 1820–21 voyage of the brigantines *Hero* and *Frederick*. These few scraps, on tattered paper and in faded ink, are all that remain of actual historic record.[2]

MEMORANDUM FOR SLOOP 'HERO'

Two Composition Rudder Braces & Two do. [ditto] pintles for hanging Ruder with the Bolts & nails for do.

200 Sheets half 18z, half 20z Best London Copper.

150 2 Inch composition nails for sheathing coppering.

150 1 Inch   ″   do for coppering.

300 Best smoothing sheathing paper.

7 Bolts best Russia Duck   6 Ditto Bear Rowens.

A Sheet chart or charts for So. America from the Equator to the highest South Latitude beyond Cape Horn. These can be got at Pattons. A sheet of charts of all the Atlantic Ocean.

FOR BOTH BRIGS

60 hhds Navy Bread; 60 bbl. Mess Beef; 40 bbl. Mess prime Pork; 4 bbl. white beans; 4 do peas; 4 do Vinegar; 10 qt. Mustard; 2 gr. chest Campay Tea; 30 do Pepper; 4½ bb. Rum; 4 bbl. Gin; 6 tt. Codfish; 2 Boxes dip candles; 1 do. sperm to be divided; 2 Boxes Soup; 8 bbl. kiln dried corn meal; 4 bbl corn; 2½ cords nut wood; ½ do. short white ash; 4 cords oak wood; 50 bus. potatoes; 3 bbl. dried apples; 5½ bbl. Rice; 12 bbl. Flour; 20 bbl. Molasses; 220 tt Coffee; 20 — 10 tt Kegs Butter; 4 Small Grind Stones; 600 feet Ceder Boards; 50 tt Boat nails; 100

assorted nails 3 spikes; 20 — 10 feet Boat oars; 40 — 16 do.;
3 — 24 feet Stering oars; 3 — 22 feet do.; 3 — 21 do.; 2 — 10 G.
Kegs lamp oil; 1 — 5 do.; 6 tt Lamp wicks; 15 Kegs white lead
ground in oil; 5 Kegs yellow paint; 6 Kegs black do; 2½ bbl.
paint oil; 20 assorted fish lines; 18 Log lines; 2 — 14 secmd
glasses; 2 — 28 do.; 2 smith bellows; 2 do. anvils; 2 do. vises; 1
doz. assorted files; 5 Boat anchors; 3 of 28; 2 Cod whale warp;
— Carpenter; Armorers Tools; 3 bbl. charco[al]; 8 wood axes; 5
boat hatchets; 125 tt sheath Leads; 25 tt bar Leads; 3 King arms;
200 hoop poles; 8 doz. skinning knives; 6 doz. steels for do.; 2½
dozen skinning knives to be made by R. Brown; 2½ doz. Beaming
knives to be made by do.; 5 bbl. Tar; 2½ bbl Teneriff wine; 5½
barrel sugar.

On these epic voyages, halfway around the planet and into
violent polar seas, the crew started in debt to the ship owner for
the purchase of warm clothing and other supplies. They worked
for shares of the catch; the captain received up to 10 percent of
the net profit, a first mate up to 5 percent, a novice deck hand
only 0.5 percent. If the sealing was good and the prices high back
home, an officer, or even a common sailor, could get rich after
just one voyage. But if the trip was a bust, everybody lost.

At sea, as on the sealing islands, their lives must have been
miserable and dangerous. The ships were wooden-hulled, usually
with copper plating. They were small, 45 to 150 tons, and vulnerable to ice; they must have bobbed like corks in the tumultuous
Drake Passage. The sea around the South Shetlands is always
near freezing, even in the summer, and the ships were unheated,
except by the galley stove, and damp. The vapor from the crew's
breaths accumulated as ice on the inner planking of the hull. In
spite of the cold, fire was a constant danger; the only light below
decks was provided by a few smoky lanterns burning sperm-
whale or elephant-seal oil. The food was a monotony of flatu-
lence-generating beans, salted meat, and maggoty biscuits. The
men relieved themselves in pots or, if it wasn't too stormy, over
the side of the ship. It was too cold to bathe. The crew worked
and slept in layer upon layer of damp woolens that reeked of

body filth, seal blood, and fat. None of the early chroniclers describe the smells, which apparently they took for granted, but they must have been overwhelming.

The sealers camped on the beach among the clamorous herds of breeding seals. It was often impossible to land on the windward beaches, where the swell slides in from the full expanse of the Pacific, and the ships would have to heave to and wait for a change in weather. Once ashore, the men had to wait for the right conditions to be picked up again. Each man was permitted to bring a small watertight bag of provisions and personal items. The sealers required only knives and clubs if they were hunting fur seals. But if they were hunting elephant seals, then oil-burning furnaces, try-pots, and staves and hoops for barrels had to be brought to the beach.

The rock-strewn sea, the fog and snow, but particularly the sudden easterly storms, constantly imperiled the sealing ships as they bobbed on anchor offshore. Robert Fildes, the leader of the Liverpool sealing fleet and master of the *Cora,* who kept a detailed journal of the summer of 1820–21, recorded that thirty-six anchors were snarled on the seafloor and had to be cut loose from the fourteen vessels in the vicinity of Livingston Island. The *Princess Charlotte,* of Calcutta, lost or broke five anchors; the *Hercules,* of Montevideo, lost four, complete with chains. But they were the lucky ones. At least six ships were wrecked in the South Shetlands during that summer. On Christmas Day 1820, the *Lady Troubridge,* of Liverpool, was wrecked off Cape Melville on the eastern tip of King George Island. Her captain, Richard Sherratt, took advantage of his misfortune; while awaiting rescue, he made an inaccurate but historically important nautical chart of the central South Shetland Islands. The exposed bay at the southeastern tip of King George Island is named after him. That same summer the first mate and ten crew members of the *Lord Melville,* of London, were shipwrecked at Esther Harbour, at the northeast tip of the island. They could not be rescued until the following spring and so became the first people to overwinter, however involuntarily, in the Antarctic.[3]

The sealing harbor of Blythe Bay, Livingston Island, was particularly infamous. It was here that Robert Fildes's *Cora* was wrecked. The stranded crew pitched a tent on the beach at nearby Desolation Island and equipped it with bunks made from wooden casks salvaged from the wreck. They lived on penguin meat, penguin eggs, and whatever supplies they could scrounge from the wreck. The ship's cat was among the survivors and soon claimed a bunk for itself. One day, wrote Fildes, "Two penguins . . . came up out of the water and took their station alongside of her [the cat] in the cask, they neither minding the people in the tent or the Cat, nor the Cat them, poor shipwrecked puss used to sit purring with their company. These penguins used to go to sea for hours and as soon as landed again would make direct for the tent and get into the cask."[4]

Not far away, at Shirreff Cove on the north shore of Livingston Island, William Smith and his sealing crew found the remains of yet another wrecked ship. It was the *San Telmo*, a Spanish galleon, which the previous summer had been part of a fleet bringing assistance to the viceroy of Peru in his battle against independence fighters. The *San Telmo*, encountering heavy seas in the Drake Passage, was dismasted and stripped of her rudder. Although she was taken in tow by one of her companions, the hawsers snapped, and on September 4, 1819, at latitude 62° S, she was abandoned. History does not record whether anybody was left aboard, but by the time Smith's men found her, there was no evidence of survivors. Perhaps because of his own close calls with storms in the Drake Passage, Smith was haunted by the wreck. He cut away the ship's anchor stock and had it made into a coffin for himself.

✳   ✳   ✳

During the breeding season, the bull fur seals, snouts elevated, imperiously gaze over their domains of beach and their harems. The beachmasters were called "wigs" by the sealers; the females were called "clapmatches." "I imagine," wrote Fildes, "the old males have been called wigs from the long shaggy hair they have about their head and shoulders, which resembles a wig, and indeed the old Gentlemen themselves look as grave as counsellors." "Clapmatch" is a reference to the frequent copulating that took place on the beach. "It was diverting to see these Old Gentlemen monopolyzing to themselfs as many females as they could, and many were the wounds we perceived when skinning of them they had receivd in Battles with others to maintain their ground."

Antarctic sealing was a carnage, a massive bloodletting, but the sealers didn't seem to care. "The time passes quickly away in the excitement of killing and flensing," one sealing captain wrote.[5] The seals were clubbed or shot to death where they lay on the beaches. Fildes observed, "Going amongst these unconcerned creatures put me in mind of Adam when surrounded by the beasts of the field at the time of the Creation. . . . I have actually seen an Old Wig endeavour to copulate with a Clapmatch . . . which had been killed with a blow to the head, and take no notice whatever of a man, tho he was taking the Skin of another Female within a foot of him."

Sometimes the harem bulls, insensate with hormone-induced territoriality, would prevent the females from escaping to the sea. "This circumstance induced the Sealers to kill the Clapmatches and leave the wigs as centinals," wrote Fildes. The pups remained near their mothers' skinned corpses and slowly starved to death. Often the harem bulls charged after the sealers and attempted to bite them, but the sealers "took the remedy of blinding them all with one eye: it was then laughable to see these old goats, planted along the beach and keeping there [sic] remaining peeper continually fixd on their seraglio of Clapmatches, while the Sailors passd . . . unheeded between their blind side and the waters edge."

Inevitably, just as the bull fur seals patrolled and defended their territories, the sealers declared their territorial rights over the same beaches, and battles erupted between the crews. William Smith wrote of the summer of 1821–22, "During the fishing season, it was with great difficulty your Memorialist maintained

Peace, between the crews of the two Nations, who were on shore." Men had come to the new Antarctic to get rich, and in an economy of unbridled slaughter, peaceful cooperation often conflicted with that goal. One of these skirmishes was recorded by Thomas Smith (no relation to William), a crew member on the schooner *Hetty,* which visited the South Shetlands during the summer of 1820–21. Smith was a mere sailor, and his below-deck record is frank and direct, in contrast to the often boastful and exaggerated accounts of the captains. After surveying several islands, the crew of the *Hetty* put ashore on a beach (not identified by Smith) "with strict orders to slaughter as many seals as we could. But on our landing, we were met by the crews of three vessels, who forbid us from taking any, claiming the beach was theirs, as they had first taken possession of it and were therefore determined to defend it as their own ground to the uttermost in their power."[6] Following the example of their competitors, the crew of the *Hetty* joined forces with those of two other vessels and staked out a beach up the coast. Their party was then the strongest in the area, and for a while the alliance worked well. But three days later, in the early morning, a crew of sealers from competing ships "walked overland to our beaches and slaughtered 8,000 seals." A battle ensued. "The leading men of our party, seeing their audacity, instantly collared their leaders to prevent them from farther prosecuting their obstinate design. This act immediately threw the parties into confusion, which resulted in a general and bloody engagement, in which many were severely injured." Only one year after the discovery of the Antarctic, men were in conflict over its resources.[7]

* * *

During the summer of 1820–21 the six vessels of the Stonington sealers were also in the South Shetlands. The tiny *Hero,* with Nathaniel Palmer in charge, served as tender, transporting the sealskins from the beaches on the south shore of Livingston Island to the brigs *Frederick* and *Hersilia* and the schooners *Express,*

Ephraim Williams, and *Freegift*. The log of the *Hero* chronicles the magnitude of the slaughter.[8]

> November 27, 1820: 1672 fur seal skins; November 29, 1207 skins; December 2, 616 skins; December 5, 906 skins; December 9, 9700 skins; December 12, 5616 skins; December 16, 6865 skins; December 19, 8229 skins; December 30, 8000 skins; January 9, 1821, 6101 skins; January 12, 2800 skins.

At first the Stonington sealers anchored at New Plymouth, a dangerously exposed bay on Snow Island. From there on November 14, Captain Pendleton dispatched the *Hero* with Palmer and a crew of four on an exploratory cruise to find a safer harbor and to scout for new sealing beaches. The next day the *Hero* arrived at the eastern edge of Deception Island, sixteen kilometers to the south, somewhere in the vicinity of the huge chinstrap penguin rookery at Bailey Head. From the sea Deception appears unremarkable: the usual piebald landscape of snow and rock. But when Palmer sailed around the southern end of the island, he found a portal into the enormous flooded caldera in its heart. Palmer and his crew were probably the first people to enter the harbor of Deception Island, a place that was to become inordinately important as a base for whaling during the next century. The island was probably named for the invisibility of the sheltered harbor from the sea. Palmer named the cut through the caldera wall Neptune's Bellows, because the wind sighed as it rushed through, and he named the notch on the eastern side of the cut Neptune's Window. From the top of that promontory, which is cloaked in fruticose lichens, Palmer claimed to have climbed the Window and to have been one of the first humans to cast eyes on the mountainous coast of the Antarctic Continent, 90 kilometers to the east. But Palmer was interested only in profit, and finding no seals on Deception, moved on.

The *Hero* continued south to the vicinity of Trinity Island, the same general area where Smith and Bransfield had journeyed ten months before. On November 16 Palmer sailed into the ice-clogged Orléans Strait, only a few kilometers from the Antarctic mainland. But seeing the channel choked with ice and observing that there were no seals, he turned back north. Had Palmer landed, he would have been the first human being to set foot on the Antarctic continent. Instead he sailed north to Greenwich Island, where he found a sheltered bay that would provide a safe anchorage for the entire fleet. On November 23, the Stonington sealers moved to this new anchorage, which they named Yankee Harbor.

They were not the only American fleet in Antarctica that summer. They were joined at Yankee Harbor by three ships, the *Huntress,* the *Huron,* and the *Cecilia,* and a company of thirty-one men from New Haven, under the command of Captain John Davis. Like the *Hero,* the *Cecilia* was a shallow-draft tender to the larger ships, bringing supplies to the sealing beaches and returning with sealskins. The *Cecilia* had not arrived in the south under her own sail but had been constructed in the Falkland Islands from precut boards brought in the hold of the *Huron.* By December 29, 1820, the New Haven fleet was having little success, having taken only 2,527 sealskins (compared to the approximately 21,000 taken by the Stonington fleet). One reason was that the sealing beaches on nearby Livingston Island were dominated by the British sealers. The captains at Yankee Harbor decided to round up their widely scattered shore parties and form a makeshift army to displace the British. On January 26, Captain Davis, in command of the *Cecilia,* joined this fleet and sailed north to Cape Shirreff, where the British sealers were camped. But by the time they arrived on the contested beaches, so few seals were left that there was no prize worth fighting for and, still without booty, the fleet dispersed. Davis and the *Cecilia* began an exploratory cruise in search of new sealing grounds, first to Snow and Smith islands, and then south across Bransfield Strait. On the night of February 5, the *Cecilia* hove to in the lee of Low Island, within sight of the mainland. The next morning she sailed south by southeast, sliding between Hoseason and Trinity islands to the vicinity of Hughes Bay, on the mainland of the Antarctic Peninsula. Davis dispatched a longboat to shore. His journal for the day reads, "Sent her on Shore to looke for Seal at 11 a.m. the Boat returned but found no signs of Seal at noon."⁹ This sparse entry is the first documented landing on the Antarctic continent, on February 6, 1821. The first human to set foot on the new continent was probably from New Haven, Connecticut, and he was wearing sealskin boots. There is no record of whose footprint it was.

*   *   *

On the evening of the same day, the *Hero,* with Palmer as master, slid quietly through the still, fog-shrouded sea between Trinity Island and Deception Island on her second exploratory cruise of the season. At midnight Palmer came on deck and took the watch. "Drifting along under easy canvass, laying to at night," Palmer recalled years later, "most of the time the mist was so dense I could not see the lookout on the forecastle. . . . I struck one bell, which brought a response that startled me; but I soon resumed my pace, turned my thoughts homewards and applied myself to the occupation of building castles in the air."¹⁰ An hour later Palmer struck two bells "that were answered by a human hand; though I could not credit my ears, and I thought I was dreaming; excepting from the screeching of the penguins, albatross, pigeons, and mother carys [pintado petrels], I was sure no living object was within leagues of the sloop." But the answering bells continued until morning. "My chief officer, who laughed at the idea of a human soul being close at hand, insisting that the sound was 'tricky' called me at seven bells during his watch, saying that voices were heard." Shortly, as the fog began to lift, the dawn revealed to the *Hero*'s starboard a frigate, and on its lee a smaller sloop, both flying the colors of Imperial Russia. Both were over ten times the weight of the forty-five-ton *Hero.* They were the *Mirnyi* and the *Vostok,* the Imperial Russian Antarctic Expedition, under the command of Admiral Thaddeus Bellingshausen.

Bellingshausen was born in 1779, the year Cook was killed in the Hawaiian Islands. Like Cook a generation before, he was a consummate navigator and explorer. Like Cook, he had been at sea since he was a child, starting as a naval cadet at age ten. From 1803 to 1806 he served as fifth officer under Captain A. J. von Kruzenstern on the first Russian voyage around the world, which proved to the czar that Russian ships could supply the sealing industry in the Kamchatka Peninsula and not have to rely on foreign vessels or on overland travel across Siberia. During May 1819, Bellingshausen was stationed at Sebastopol in the Crimea, when he was summoned to St. Petersburg by Czar Alexander I. The overland journey took him more than a month. Bellingshausen was given six weeks to organize and lead the first Russian Antarctic exploration. His two ships, constructed of unseasoned pine boards, were ill suited for polar sailing. The *Vostok*, at 600 tons and 39 meters long, was a copper-sheathed man-of-war equipped with twenty-eight cannons; the *Mirnyi*, at 530 tons, had no copper plating, but her hull was sealed with pitch-soaked canvas. Bellingshausen's second in command, and the captain of the *Mirnyi*, was Mikhail Lazarev, as capable a seaman as Cook's Tobias Furneaux. The crew consisted of 189 officers and sailors, all carefully chosen. Each crew member was required to be under thirty-five years old and to have at least two skills. The ships set sail from Kronstadt on July 16, 1819.

In many ways Bellingshausen's exploration continued where Cook's left off. Cook's voyage took three years; Bellingshausen's, two. Bellingshausen deliberately chose to explore sectors of the Southern Ocean that complemented but did not duplicate Cook's voyage. Cook had managed to traverse a total of 24 degrees of longitude south of the sixtieth parallel; Bellingshausen, in the years after the Little Ice Age, sailed 42 degrees longitude south of that meridian. At the beginning of the austral summer of 1819–20, he sailed south-southeast from Rio de Janeiro to South Georgia. Although forty-four years earlier Cook had explored the northern shore of the island, and it had been visited many times by sealers, there were very few charts of the island. Bellingshausen. mapped the southern shore. He then proceeded to the South Sandwich Islands, where he discovered and circumnavigated three more islands in the group, proving that the archipelago was not the northern tip of the Antarctic continent. The ships proceeded southeast into the Indian Ocean. On January 27, 1820, the *Vostok* and the *Miryni* crossed the Antarctic Circle, the first vessels to do so since Cook's. On January 28 to 29 and again on February 5, Bellingshausen's expedition was within sight of the mainland, but the weather was so foul that vision and reality blended once more.[11] His journal on February 17 conservatively states:

> On the horizon towards the south a vivid brightness appeared which indicated dense pack ice. Towards midday the mist cleared and dry snow, which had fallen at times, ceased, but the sky remained cloudy. . . . The farther we proceeded, the more dense became the ice until about 3.15 p.m., when we observed a great many large high flat-topped icebergs, surrounded by small broken ice, in places piled up high. The ice towards the south-south-west adjoined the high icebergs which were stationary. Its edge was perpendicular and formed into little coves, whilst the surface sloped upwards towards the south to a distance so far that its end was out of sight even from the mast-head.[12]

Bellingshausen, at 69° 25′ S, 1° 11′ W, somewhere in the vicinity of the Ritscher Upland, was probably gazing on the Antarctic continent. But judging from his descriptions, he didn't know it. In a sea where ice seems to be snow-covered land, the cautious Bellingshausen assumed that the mountains on the southern horizon were merely massive icebergs.

Bellingshausen spent the first winter in Australia, and after refurbishing his ships, explored the South Pacific, including the reef-strewn Tuamotu Archipelago. While at anchor in Sydney Harbor, he heard reports of the discovery of the South Shetland Islands and the sealing activities there, and he resolved to explore

the area. Come spring, he sailed across the ice-choked mouth of
the Ross Sea, and on January 22, 1821, reached the farthest south
of his voyage (69° 53′ S; 92° 19′ W, still about 160 kilometers
north of Cook's deepest penetration). The next day Bellingshausen wrote, "We saw a black patch through the haze to the east-
north-east. I knew from my first look through the telescope that
land was in sight." But reality and illusion still haunted the explorers; "The officers, when they looked at it through glasses,
were divided in their opinion." A few minutes later,

> The sun coming out from behind the clouds lit up the place and to
> our satisfaction we were able to assure ourselves that what we saw
> was land covered with snow. Only some rocks and cliffs, where
> the snow could not hold, showed up black. Words cannot describe
> the delight which appeared on all our faces at the cry of 'Land!
> Land!' Our joy was not surprising, after our long and monotonous
> voyage, amidst unceasing dangers from ice, snow, rain, sleet and
> fog.[13]

What they saw was an island; Bellingshausen christened it Peter I
Island, after the czar who opened Russia to the West. Unaware
that he had seen the Antarctic mainland the year before, Bellingshausen believed that the island was the first land observed south
of the Antarctic Circle. On January 29 land was again sighted.
Bellingshausen named it Alexander I Land and thought that he
had at last seen the mainland. Not until this century did geographers learn that it was an island, although fused to the mainland by an ice shelf that spans the thirty-two-kilometer-wide
George VI Sound.

A week later, on the ice-fogged sea near Trinity Island, the
Imperial Russian Antarctic Expedition encountered the tiny
Hero. Bellingshausen dispatched a longboat to pick up Palmer.
The Russian admiral, in full dress uniform, was regally seated at
his desk in his cabin aboard the *Vostok* when Palmer, wearing a
sealskin coat and boots, strode aboard. Years later Palmer recalled, "I gave him an account of my voyage, tonnage of sloop,

number of men and general detail, when he said: 'How far south have you been?' I gave him the latitude and longitude of my lowest point, and told what I had discovered. He arose much agitated, begging that I would produce my log book and chart, with which request I complied, and a boat was sent for it." Over lunch Palmer and Bellingshausen discussed the South Shetlands and in particular Yankee Harbor, where the rest of the Stonington fleet was sheltered. And now Palmer's memory of the encounter becomes florid.

When the log book and chart were laid upon the table he examined them carefully without comment, then rose from his seat saying: "What do I see and what do I hear from a boy in his teens [Palmer was in fact twenty-one]: that he is commander of a tiny boat the size of a launch of my frigate, has pushed his way towards the pole through storm and ice and sought the point I in command of one of the best appointed fleets at the disposal of my august master have for three [actually two] long, weary, anxious years, searched day and night for. . . . My grief is your joy. Wear your laurels."[14]

Palmer's tale was probably exaggerated. In Bellingshausen's journal the report of the encounter is perfunctory; the man who had first laid eyes on land south of the Antarctic Circle was probably nonplussed by the braggings of the grubby twenty-one-year-old sealer. His journal entry simply states, "Mr. Palmer arrived in our boat and informed us that he had been there for four months' sealing."[15] Bellingshausen seems to have been shocked at the rapacity of the sealers in the South Shetlands. "Mr. Palmer told me that . . . Captain Smith, the discoverer of New Shetland, was on the brig *Williams,* that he had succeeded in killing as many as 60,000 seals, whilst the whole fleet of sealers had killed 80,000." One year after the discovery of Antarctica, Bellingshausen voiced what may have been the first words of concern for the Antarctic environment. "As other sealers also were competing in the destruction of the seals there could be no doubt that

round the South Shetland Islands just as at South Georgia and
Macquarie Islands the number of these sea animals will rapidly decrease."

✳   ✳   ✳

By 1822, fur seals were already becoming rare in the South Shetlands. "This valuable animal, the fur-seal," wrote James Weddell,
"might, by a law similar to that which restrains fishermen in the
size of the mesh of their net, have been spared to render annually
100,000 furs, for many years to come. This would have followed
from not killing the mothers till the young were able to take to
the water; and even then, only those which appeared to be old,
together with a proportion of the males, thereby diminishing their
total number, but in slow progression."[16] Weddell, the son of a
Scottish upholsterer and an English Quaker, made two sealing
trips to Antarctica, including the South Shetland Islands, between
1822 and 1824. Like William Smith, Weddell was primarily a
merchant and a sealer, but he was also a man of great natural
curiosity. In 1823, as captain of the brigantine *Jane*, he journeyed
farther south than any other man (74° 15' — a record that was
not broken until 1841, when James Ross reached 78° S), in the
sea east of the Antarctic Peninsula that is now named after him.
Two years later Weddell published *A Voyage Toward the South
Pole,* a celebrated book on his voyages. In it he gave the first
description of the seal that now bears his name, but the accom-
panying drawing is of a pinheaded animal that looks more like a
toothed grub than a seal. He also indulged a romantic, but false,
notion that was prevalent among the sealers of his time. Referring
to elephant seals, he wrote, "If the skull be indented in the killing
of a female with young, the indentation is found also upon the
skull of the young. This sympathy, which has been denied with
regard to the human species by some physiologists, evidently
exists in the economy of this animal."[17]

Weddell was popular among his crews for his liberal allow-
ances of rum: three glasses per day per sailor. This may have

contributed to one of the more bizarre observations made in the Antarctic, the sighting of a mermaid somewhere in the South Shetlands. Weddell wrote that one of his sailors

> . . . had gone to bed, and about 10 o'clock he heard a noise resembling human cries. . . . He walked the beach a few steps. On searching around, he saw an object lying on a rock, a dozen yards from the shore, at which he was somewhat frightened. The face and shoulders appeared of human form, and of a reddish colour; over the shoulders hung long green hair; the tail resembled that of a seal, but the extremities of the arms he could not see distinctly. The creature continued to make a musical noise while he gazed about two minutes, and on perceiving him it disappeared in an instant.

The sailor was of course doubted. But, Weddell continued, "To add weight to his testimony (being a Catholic), he made a cross on the sand, which he kissed in form of making oath to the truth of the statement."[18] The vision was perhaps a seal with algae on its head, and from the vantage of our time, it seems incredible that Weddell included the tale in his book. But his acceptance of the vision reveals the marvelous expectations of those first travelers to the Antarctic. It was an unknown land, endowed with dark mysteries, in which anything was possible. After all, it was only three years earlier that Smith and Bransfield were under orders to contact and appease any Antarctic natives. Even today, 170 years later, Antarctica is the least known of the seven continents.

✳   ✳   ✳

By 1825, only six years after the European discovery of Antarctica, the Antarctic fur seal was nearly extinct in the South Shetland Islands, as it had become on South Georgia Island, and as had the South American fur seal in Patagonia and the islands of the eastern Pacific. Weddell estimated that 320,000 sealskins were taken from the South Shetlands during 1821 and 1822 (this figure does not count the 100,000 suckling pups that starved to death next to their mother's corpses). Bellingshausen's fears were realized; the sealers had extinguished the very resource that sustained them. Fortunately, a few survived, and later in the century the fur seals began to return to the islands, but a short flurry of sealing expeditions in the 1870s and 1880s exterminated them again. To this day, on some beaches above the mark of the scouring tides, heaps of seal bones from that era can still be found. Today descendants of the few surviving fur seals are returning once more to their natal beaches in the South Shetlands. The species, at least for now, is fully protected according to the terms of the Convention for the Conservation of Antarctic Seals. In recent years a few Antarctic fur seals, both males and females, have returned to Admiralty Bay. They do not breed in the bay but arrive late in the season to molt and rest after the strain of procreation.

The meager records left by the first voyagers to Antarctica have invoked a centuries-long and fruitless retrospective debate among the British, Americans, and (lately) Russians as to which nation's hero should claim the discovery of the continent. Each nation presents arguments that are devoid of the necessary proof. It seems irrelevant to these claimants that the first British and Americans came to despoil and plunder Antarctica's natural resources, and after six years left the fur seal almost extinct.

In 1821 Nathaniel Palmer's adventures were only beginning. His was to be a hero's life. The next summer he and the English sealer William Powell discovered the South Orkney Islands. Later Palmer ran the Spanish blockade of Cartagena, bringing supplies and troops to Simón Bolívar, and in the 1830s Palmer was instrumental in the invention and design of the famous clipper ships, of which he built and owned several. He grew rich trading with China.

After his return to Russia, Bellingshausen fought in the Russo-Turkish war and later became military commander of Kronstadt. The news of his discoveries of land south of the Antarctic Circle thrilled Europe. Ludwig van Beethoven, deaf and angry in his last years, was reading the accounts of Bellingshausen's voyage when he composed his *Missa Solemnis* and the Galitzin quartets. But it wasn't until 1945 that Bellingshausen's journal was translated into English. Peter I Island was not observed again by human eyes until 1910, ninety years after its discovery, when the Norwegians landed there and annexed it to their own Antarctic territory. The Russians didn't return to the Antarctic until 1946, when they started whaling in the Southern Ocean.

William Smith returned to Portsmouth in September 1821, but upon his arrival the discoverer of Antarctica learned that his partners had gone bankrupt. The *Williams* was sold and eventually was used once again to carry coal among the British Isles. It was the purpose for which she was originally built, but an ignominious end for the ship that was to Antarctica what Columbus's *Santa Maria* was to the Americas. And Smith, the last human being who would ever discover a continent, slid into poverty and obscurity. His explorations are not taught in history classes; his name does not appear in most encyclopedias. He received a license to pilot ships locally in England and became master of a whaling ship in 1827, but he was fired three years later. In 1838 Smith applied for residence in an almshouse but was rejected for being too young. A year later his pilot's license was revoked because his skills were "greatly reduced by age and infirmity." Smith petitioned the British Admiralty for remuneration for his discovery of Antarctica. Ironically, the request was referred to Captain Shirreff, the very man who had at first ridiculed Smith's discovery, and was denied. Smith died a pauper. It is not known whether he was buried in his Antarctic coffin.

10
The Tern and the Whale

First to the Sirens ye shall come, that taint
The Minds of all men whom they can acquaint
With their attractions.

— HOMER

THE SECOND TEMPEST broke this morning, and the day is fits of gust and bluster, a respite from the sustained shrieking of the two previous days. The cabin fever has also broken, and during the long hours of the late Antarctic summer, when the sun lopes along the northern horizon and the light is angular and clean, I walk behind the station, high up on the tumbling slope, beyond the ruined wooden huts and the four wind-etched crosses, over the first undulating hills and into the swale beyond. The ground is swollen and oozing. The wet rocks reveal every nuance of strike and dip, embedded pebbles, the intrusions of ancient magmas — details that are invisible when they are dry. The ephemeral stream that drains the glaciers and snowfields is now a torrent, depositing a plume of silt in the bay. To the southwest the cloud-wind is streaming over Pilot Mountain, sharply etched in the blasting sunlight. The vapor-bearing air flexes as it passes over the smooth, broad mountain, like wind over an airplane's wing, and as the air rushes off into the void over the bay, the clouds evaporate, pink to nothing.

Two curious south polar skuas follow me, looping and protesting my intrusion. They must have chicks nearby. They fly at my head low and fast from the direction of the sun. Birds of darkness, they hide in the light. I hold my hat above my head as a decoy, and they slap it with their legs.

On the next slope, in a cluster of frost-shattered boulders I find a single chinstrap penguin, eyes closed and feathers fluffed against the wind. It is molting and has climbed into the shelter of the rock to get away from the freezing wind. During molt the skin of penguins becomes engorged with blood, and they are particularly vulnerable to the cold. Beyond the penguin are the nests of Antarctic terns. Terns form small colonies and will collectively defend their progeny, but they are not gregarious and build their nests far apart. Several pairs have chosen to nest in the shelter of this valley. Now the skuas fly in tighter. Cunning predators that use surprise in their own hunting, they know I may flush a baby tern from the rocks. But the adult terns rise to defend. Their heads are black-capped, and the sharp, air-cutting edge of their wings is black, but their bellies are as white as snow and invisible against the sky. The terns harry the skuas, monotonously screaming, always approaching their bigger enemies from above. They also know how to attack from the direction of the blinding sun, and just above their quarry, they flatten their wings against the sustaining airflow and stall, diving onto the slower birds with claw and bill. The skuas veer and air-twist, but to no avail; the two birds tumble and separate, and the attack begins again.

Antarctic terns, like most terns everywhere, scrape shallow depressions in the bare earth for their nests. Their mottled eggs mimic earth and sand, and the dowdy chicks, when frightened, freeze and become stones. I search among the jumbled shards of rock for the chicks, carefully placing each footfall so as not to flush a chick or step on an egg. Judging from the scattered down and the streaks of white scat, the chicks are close, perhaps centimeters away. But to me they are invisible. The skuas, equally frustrated, peel off high overhead into the turbulent, invisible gusts, pretending disinterest, but hounded nevertheless by the angry terns. They are defeated; these are no easy, land-bound penguins.

And then I see a chick, huddled behind a windbreak of rock. It is the exact color and texture of the earth itself. Its instincts tell it not to move, not to blink, not to breathe. Evidently it has been watching me for some time. I pause for a moment. This indeed is a great accomplishment.

Two species of terns visit the Antarctic Peninsula during the summer. The Antarctic and the Arctic tern are very similar in appearance — about the same size and with similar plumage — but their reproductive behaviors are poles apart. The Antarctic tern breeds here and during the winter migrates to warmer seas off the coasts of South America and South Africa. The Arctic tern breeds in the Arctic antipode and is a nonbreeding visitor to Antarctica during the austral summer. Its migration, 18,000 kilometers from the Arctic to the Antarctic and back, is unmatched among animals; it simply couldn't fly much farther on planet Earth. There are several breeding populations of Arctic terns, but the one that frequents the Antarctic Peninsula is probably from the lands that fringe the North Atlantic sector of the Arctic Ocean. Starting in October, these birds migrate south via Europe and Africa, then cross the Southern Ocean to the edge of the Antarctic pack ice. By the time they arrive in Antarctica, they are in nonbreeding plumage: slightly white-pated, the edges of their wings a muted gray. Feeding at sea on krill and fish, they take advantage of the East Wind Drift and follow the gyre of the Weddell Sea west and north before returning to the Northern Hemisphere in March. Arctic terns do not assume their breeding plumage until March, about the time that the resident Antarctic terns begin to lose theirs. The Antarctic and Arctic terns are therefore at opposite phases of their reproductive cycles during the austral summer.

This chick will need parental care before dusk; not wanting to renew the skua's attention, I steal quietly away. I walk to a nearby promontory overlooking the white-capped bay. One of the skuas is perched on the highest point, looking like a contented hen. With the seal placentas cleaned up, baby penguins fledged and at sea, and autumn impending, times will soon be hard for these birds. She growls at me, sounding like a slurring, drunken crow, and flies off, low and invisible, brown and white, against brown and gray stones; she too is earth and rock. Sheets of wind tumble across the bay, tousling it like a field of wheat. The terns flying to the bay to feed have to loft over the updrafts of this wind ridge, and they are startled to see me again. They pause overhead and, one by one, hover on bent wing and scrutinize me, turning their heads from side to side. At last, deciding that I am innocuous, they slide into the undulating bay gusts, hill by hill, to the black beach and then to the sea. On the beach far below there are a few solitary chinstrap and gentoo penguins. Occasionally, in the calm between gusts of wind, I hear one yelping, a vestige of breeding times now past, but dimly, hormonally, remembered.

And beyond the beach, in the black, choppy sea in front of Ullman Spur, are five humpback whales, three adults and two juveniles. Their spouts are tattered by the wind, and from this height they are difficult to distinguish from the agitated water. Antarctic prions, perhaps feeding on droplets of exhaled oil, reel and dip in the confused water where the humpbacks surface. I run to the top of the low hill by the abandoned British station to get a closer view of the whales, and wait. Nothing. And then I see the huge white pectorals, reflected up from the blue like an underwater incandescence, the vapor of a breath, a smooth, broad back arcing in the sea, and the small fatty nubbin of a dorsal fin. Finally the flukes, tattered and piebald as if spattered with white paint, rise from the sea and are held in the air momentarily, a gesture to the clouds and mountains, before sliding into the deep. The whales are sounding to feed on krill, themselves rising at this hour to greet the darkness. They are gone several cold minutes and then appear again; I can never quite anticipate where. They surface in front of my promontory now. The largest rolls on its

pleated side and lazily waves its flipper in the air, then slaps the
water. The babies, all gangly flukes and fins, stay close to the
adults, but one, in an occasional exuberance, breaches. As if in
slow motion, its plump body — of a scale beyond anything human — rises from the sea and for a long instant, turning, is poised
in air.

✳   ✳   ✳

Today it is rare to see a whale in the Southern Ocean. Whalers
have reduced the right, humpback, fin, and blue whales to vestigial numbers. But the first travelers to Antarctica observed large
numbers of these animals. James Ross, who sailed through these
waters in December 1841, wrote:

> We observed a very great number of the largest-sized black [right]
> whales, so tame that they allowed the ship sometimes almost to
> touch them before they would get out of the way; so that any
> number of ships might procure a cargo of oil in a short time. Thus
> within ten days after leaving the Falkland Islands, we had discov-
> ered not only new land, but a valuable whale-fishery well worth
> the attention of our enterprising merchants, less than six hundred
> miles from one of our own possessions.[1]

The next month he wrote, "In the course of the day a great
number of whales were observed; thirty were counted at one time
in various directions."[2] Fifty-seven years later Frederick Cook
sailed through these waters on the *Belgica*. One of the recurring
themes of his eloquent journal was the abundance of great
whales. He wrote, "In the waters are large numbers of finback
whales which, with the seals, will in the near future offer a new
industry."[3] Unfortunately, the prophecies of Ross and Cook were
all too soon fulfilled.

Humpbacks are baleen, or mysticete, whales, of which there
are six species in the Antarctic.[4] These include the four species of
rorquals — the blue, fin, sei, and minke whales — different-sized

variations on a theme of streamlined, fusiform bodies, large flippers, and small, fatty dorsal fins. The humpback differs from the
rorquals in having inordinately large pectoral flippers, up to five
meters long, which may be an adaptation for tight maneuvering
in shallow coastal waters. The sixth baleen whale, the right
whale, is also found in these coastal waters. It has no dorsal fin,
and a huge calloused head that takes up one-quarter of its body.

It is the filtering plates of baleen that differentiate the mysticetes from the toothed whales, or odontocetes, of which five
species are found in the Antarctic: the sperm whale, the orca, the
southern bottlenose whale, the long-finned pilot whale, and
Layard's beaked whale.[5] Paleontologists believe that about fifty
million years ago odontocete whales evolved from four-legged
terrestrial ancestors, and about 20 to 25 million years later the
mysticete whales branched off from the toothed whales. Until
recently no fossil remains of this transition from tooth to baleen
had been found. But in 1989 the jaw of an ancestor of the baleen
whale, 40 million years old, was found on Seymour Island. It had
deeply lobed teeth, in appearance almost identical to those of a
crabeater seal, and no doubt they were used in the same manner
to filter food from the sea. So whales have been living, and dying,
in the sea for perhaps as long as 50 million years. The seafloor is
a record of the antiquity and abundance of whales, and, particularly along their migration routes, is littered with whale bones.
One dredge of the ocean bottom made by the pioneering research
vessel *Challenger* (1782–1786) recovered, in addition to over a
thousand sharks' teeth, sixty whale ear bones in a single haul.

The mysticetes are the ecological, and in many ways behavioral, equivalent of the grazing antelopes of the East African
savanna. They graze on zooplankton, which in the Southern
Ocean is almost invariably krill, and, to a lesser extent, on copepods. Their mouths are fretted with sieves of baleen hanging in
two rows from the palate. This is the misnomered whalebone,
once widely used in commerce. In fact baleen is not bone at all; it
is made of keratin, the same protein that is in claws, fingernails,

and the horns of rhinoceroses. The humpbacks and rorquals feed by lunging, open-mouthed, into the planktonic sea. Their jaws can open as much as ninety degrees, and the pleated throat expands like a balloon; a large blue whale can gulp as much as 60,000 kilograms of food and water. When the mouth closes, the huge tongue, acting like a piston, pushes the tons of water through the baleen plates, sieving it of krill and other morsels. The right whale and sometimes the sei, employs a different method of feeding, called skimming, in which it swims, open-mouthed, through swarms of copepods and krill. The forward motion rather than the tongue pushes the water through the baleen.

The coarseness and bristle size of baleen vary according to species, thus differentiating their niches in a sea where the principal food is one or two species. The baleen of the copepod-eating right whale is denser than that of the humpbacks and rorquals, and the plates are so long, up to two meters, that they have to fold when the whole closes its mouth. Blue whales, in spite of their size, specialize in shallow swarms of adolescent krill; fin whales feed on deeper swarms of adult krill. To digest this huge amount of chitin, fat, and protein, most baleen whales have a three-chambered stomach, the anterior chamber of which is highly muscular and equipped with grinding plates, followed by a folded intestine five to six times the whale's body length.

The baleen whales of Antarctica live a relatively short time — at best only twenty-five or thirty years — and grow fast, but they reproduce slowly. They spend only the summer in the Antarctic, passing the austral winter in tropical waters to the north, pushing their huge bulks one-quarter of the way around the world and then returning.

During the dark winter, there is nothing for baleen whales to eat in the ice-smothered Antarctic sea. In the still water under the pack, there is no wind-shuffling of the sea surface, no stirring of the nutritious sediments from the shallow ocean bottom. The krill cling to the underside of the waxing and darkling ice, making do on the film of diatoms that taint and blotch the pack.

And so the whales, fattened by the feast of the Antarctic summer, swim north, through the howling fifties and the roaring forties, to more pacific seas. The humpbacks migrate the farthest, up the west coast of South America, past the coastal deserts of Chile and Peru, to the equatorial waters off the coast of Ecuador, a journey of about 6,500 kilometers. The right whales migrate only as far as the sheltered bays of Patagonia. The blue, fin, and sei whales migrate a shorter distance than the humpbacks, but farther than the right whales. The journey is remarkably fast; swimming day and night at speeds of up to twenty-four kilometers per hour, a sleek blue whale may reach its winter habitat in as little as fifteen days. One sei whale tagged by researchers swam 3,550 kilometers in only ten days. The humpbacks are less streamlined and undoubtedly much slower.

The migrations of the great whales, during opposite months in the Northern and the Southern Hemisphere, have resulted in two disjunctive populations, north and south. The long-distance migrators, such as the humpbacks and blue whales, visit equatorial waters six months apart and therefore probably haven't interbred, at least since the Pleistocene glaciations. The short-distance migrators, such as the right and sei whales, don't even come close to their counterparts in the other hemisphere. Along the way there may be some opportunities for feeding, particularly in the upwelling Humboldt Current that brings boom or bust to the Peruvian anchovy fishery every year. But throughout most of their migration and winter residence in the sterile waters of the north, the baleen whales fast.

The Antarctic tern makes an equivalent migration, to the temperate south Atlantic, and the Arctic tern migrates practically from pole to pole, but these animals weigh only a few ounces. A large blue whale weighs more than a million terns. While terns slice through cold air, the whales slide through frigid water (in fact, often much warmer than the air), past upwelling and berg, on their global peregrinations. But the energetics of whales and of terns are poles apart. The tern, with a wingspan of only sixty-one centimeters, has a ratio of surface area to volume that is huge compared to that of a whale; its heart flutters like a butterfly's wing beat, and it burns far more energy per ounce than the whale. And all through its flight, it dips into the sea to capture fish or invertebrates. The baleen whale, by contrast, is huge and blub-bered, its surface area small relative to its bulk, and its heart beats, in a slow swim, only eight times a minute.

No one knows why whales make these epic, starved journeys. Although Antarctic whales do not calve in polar waters, there is no obvious reason why they couldn't. Bowhead and beluga whales, for example, both give birth in frigid Arctic waters. The migrations may be prehistoric accidents, behavioral remnants of the Pleistocene ice ages, when the waters of the middle latitudes were considerably colder, and more productive, than they are today.

Animal migration remains one of the abiding mysteries of biol-ogy. Little is known of how the whales navigate or whether they return to particular feeding or breeding grounds at either end. The ocean echoes like a hollow room, and just as a blind person can sense the presence of an object by the subtle sound waves bouncing off it, the mysticetes may be able to interpret nuances in the background noises of the sea. Or they may sense thermal discontinuities and maintain their course within currents. Per-haps, like Eratosthenes, whales measure the orientation of the sun, which shifts higher in the sky as they swim north, hinting of decreasing latitude.

Or they may use magnets. There is mounting evidence that migrating species as disparate as honeybees, homing pigeons, and humpback whales are able to sense and follow the direction of the Earth's magnetic field. The observation that whale strandings seldom occur along shorelines that run parallel to the Earth's magnetic field supports the hypothesis. These animals use minute particles of magnetite, an oxide of iron. Magnetite is a mixture of ferric and ferrous iron, which have opposite electron spins and therefore create a miniature magnetic dipole that naturally tends to align itself with the Earth's magnetic field. Crystals of magnetite, ranging from 0.05 to 0.20 millimeters, are commonly found in the brains of migrating animals, including the cerebellums of humpback whales. How the animals perceive the orientation of the magnetite particles is unknown. One hypothesis is that the magnetite particles affect the electrical resistance of cell membranes and concomitantly the frequency of electrical discharges, but this mechanism has not been proven.

But the Earth's magnetic field is not constant; the magnetic poles, those places where a compass needle points down, migrate ten to fifteen kilometers per year, as the Earth's molten iron outer core, which generates the magnetic field, shifts. Since 1908, when Edgeworth David, Douglas Mawson and Alistair Mackay first reached the south magnetic pole, it has shifted from Victoria Land on the Antarctic Plateau to a point in the Dumont d'Urville Sea, 800 kilometers away. Migrating animals have had to adjust their compasses to the gradual changes over the millennia. They are tuned to changes deep in the inanimate heart of the planet.

During their migratory journeys the whales sing. Once I swam with humpbacks on their calving grounds near the Galápagos Islands. These may have been the same whales that visit Admiralty Bay. As soon as I put my head in the water, I heard their songs, seemingly coming from all directions. In the clear, warm water the songs seemed unearthly — a coloratura of trumpets, low whoops, noises like creaking hinges. The low notes were beyond the range of my hearing; I sensed them as a reverberation in my chest and lungs and cranium. These were certainly like no other animal sounds I had ever heard. Early mariners, in their quiet sailing ships, found the mysterious songs haunting as they seeped through the wooden planking from the dark sea below. Were they the sounds of sirens? Angels? The sirens that cried to Ulysses were probably humpback whales. Indeed, the Sirenusian Isles, off Capri in the Bay of Naples, may have been a calving area for whales. Sadly, the humpback whale is now extinct·in the Mediterranean. Some of the earliest records of incorporate voices emanating from the sea are found in Irish legends. The seafaring Maeldune, in the eighth century, drifted to enchanted islands. "And as they went on," he wrote, "they heard in the north-east a great shout and what was like the singing of psalms."[6] When I have slipped into the water in Admiralty Bay and, holding my breath, listened, I have heard just about everything but whales: the popping sounds of crustaceans, the descending Doppler gongs of Weddell seals, and the ping and snap of melting icebergs. Alas, there have been no whale songs, let alone psalms.

Humpback whale songs are truly marvelous productions. The fundamental units are notes, which are arranged into subphrases, then phrases, which are in their turn merged into themes. These are then recited, and repeated, in no particular order. The subphrases often rhyme, in what may be a mnemonic adaptation for remembering the songs, although the rhymes are not obvious to a human listener. Like a skilled opera singer, the whale steals breaths without interrupting his song; the longest recorded humpback song lasted twenty-one hours without a break, but they may last much longer. Although whale songs conform to certain rules of composition, they are not the invariable croakings of animal instinct, like the cries of the terns above. What distinguishes humpback whale songs is that they progressively change from year to year, and all singers within acoustic range simultaneously adopt the new versions. In all of nature, only one other nonhuman species is known to modify its songs in this manner: the yellow-rumped cacique, a highly vocal, communal-nesting bird of the Amazon forest. The song of the cacique can perhaps best be described as sounding like water gurgling over stones. I have a friend, a rubber tapper in a remote area of the Brazilian Amazon, who built his homestead under a tree dripping with the hanging nests of yellow-rumped caciques. "Their melody," he explains, "casts a special spell over the forest, especially at dawn and at dusk. For this reason I have the best house on the Amazon River."

Why do the whales sing? No one knows, but there are interesting hypotheses. Like male canaries, male humpbacks may sing to attract, and seduce, females, which, also like canaries, may tend to favor those suitors with the largest vocal repertoires. The singers are mainly males escorting females with calves. Whale songs certainly communicate information, but the long process of correlating voice with behavior is just beginning, and the changing of songs over the years adds to the confusion. In a way it is like burning one's library every few years. In the still and quiet depths, away from the chop and torment of the sea surface, whale songs may travel hundreds of kilometers, and a pod, linked acoustically, may be just as extensive, the virile males declaring their vigor across hundreds of kilometers. During migration, therefore, the songs may serve as beacons, animate lighthouses in a trackless sea.

*   *   *

As the northward migration begins, the males show a marked increase in testicular size and sperm production. In the equatorial seas, the mysticete whales of the Antarctic procreate. Although cetologists have been dissecting mysticete whales on the flensing platform for most of this century, few have observed living rorquals making love. The whales are widely dispersed and seldom seen in the warm seas, and of course it takes place under water. The actual act of intercourse lasts less than a minute, perhaps only a few seconds. However dispassionate it seems, sex among the mysticetes is a Brobdingnagian affair. The pointed penis of a large bull blue whale may be three meters long and is prehensile, enabling adroit insertion on the high seas. The pelvic bones of whales, which float freely in soft tissue, once were thought to be vestigial and without use; it is now known that they anchor the penis during copulation. Coitus is usually belly to belly, with the whales rolled on their sides, although some observers have described humpbacks mating as they rise from the depths and make a partial breach. The rorquals are promiscuous (although blue whales are now so rare that a female may have a hard time finding *any* mate). Because the males do not gather and defend harems, there is little advantage to large male size. The females are as large, or larger, than the males.

The males have it easy; a single spermatozoan — one cell — is all they contribute to each of the calves that they sire. But getting that sperm to the egg, in the face of competition from other lovers, is no easy task. The males compete with copious amounts of semen — gallons, probably — in an attempt to wash away the sperm of the female's previous consorts. To manufacture this prodigious output, each testicle of a right whale weighs about 2,000 kilograms — the weight of a medium-sized truck. Since the males have no certainty as to the paternity of the calves, they invest nothing in their raising.

The females, relying on stored fat and protein from the summer's feast must gestate the fetus over the course of a year while traveling as much as 6,500 kilometers to the Antarctic and 6,500 kilometers back. In eleven months the zygote of a blue whale grows from a single cell to a length of eight meters and a birth weight of 2,000 kilograms. The growth is not constant. The fetus's development is delayed for the first months while the mother is fasting, but when she reaches the rich Antarctic feeding grounds, the fetus begins an exponential growth phase, doubling its weight every few days. It is born soon after she returns to the warm tropical seas where it was conceived the year before. The explosive growth rate continues after the calf is born. In a single day a blue whale calf consumes about 100 kilograms of milk, and the newborn doubles its weight in only seven days. Whale milk, which has the consistency of thick cream, is 36 percent fat, 13 percent protein.[7] The two breasts are located in grooves on the belly, on either side of the genital cleft. Since suckling is impractical in the water, the teats are equipped with powerful muscles that squirt the milk into the infant's mouth.

In November the blue and minke whales are the first to arrive in the Antarctic, and the first among them are the pregnant females. They are soon followed by the males, and last, by the mothers with suckling calves. The humpbacks follow in December, again with pregnant females in the vanguard. The fin whales don't arrive in large numbers until January, and many of the sei and right whales never arrive, but instead linger in the vicinity of the Convergence or in the waters north of it.

The period of maternal dependency is short, and the calves will be weaned before summer ends. A rorqual conceives, gestates, nurses, and weans a calf every two years, and by the time a lactating female returns to the tropical seas of winter, she is in estrus once again, and the long, debilitating pregnancy begins anew.

The austral summer is the time of gorging and fattening. Life is unimaginably abundant — and cheap — in the Southern Ocean during the summer. An adult blue whale eats as much as 3,600 kilograms of krill per day. The pregnant females, which not only are nourishing a fetus, but must also set aside reserves for the long period of lactation ahead, gain the most weight. A rorqual typically consumes 3 to 4 percent of its body weight per day, and in the course of an Antarctic summer can double its mass. Most of this weight gain is fat, the main constituent of blubber, and fat also insinuates itself into muscle, bone, and, especially, the tongue. Krill, particularly the gravid females, are very nutritious, up to 11 percent protein and 7 percent fats. By March, with autumn advancing, the pregnant females are the last whales to leave on the journey northward.

Although the Antarctic baleen whales are gregarious (one Antarctic pod of minkes consisted of about a thousand animals), there is only rudimentary social organization on the feeding grounds. Pods fission and coalesce according to the swarms of krill, the currents, and the weather. Rorquals frequently feed in flotillas, advancing on a krill swarm, sounding, and breathing in synchrony; right whales have been observed skimming in chevrons, like an echelon of flying geese. But the humpbacks are the most innovative plankton eaters. They weave circular nets of exhaled bubbles, which confine the krill as effectively as any nylon net, and then they rise, open-mouthed, through the center. In an environment where there is only air and water, air is a raw material that is wrought into a new shape and use, and this, I believe, is genuine tool-making.

✳   ✳   ✳

A few weeks ago a minke whale, the smallest of the rorquals, visited the bay. We were in the Zodiac, retrieving the amphipod trap from the opposite side of the bay, when the minke slid through the swell beside us. We turned the Zodiac toward the surprised whale to get a closer look. This whale was not one of the lazy humpbacks that loiter in the bay, indifferent to humans. Today's humpbacks have no memory of the hunt. But minkes are still hunted, and this little whale panicked. It swam to the deep center of the bay and we followed. I could see the dimples on its face, the thick lips, the sheet of baleen hanging from its palate like a frozen aurora, its waxy eye, rolling with intense frightened curiosity, the lovely gray chevron on its flank. Its smooth back undulated as if it were one with the swell. I could hear its urgent breathing and smell its breath, oily and halitotic. Then the whale sounded, and the bay was suddenly still, until it surfaced, unpredictably, a hundred yards away. The chase was becoming harassment, and we stopped. The minke surged toward the mouth of the bay and disappeared into the lumpy horizon of sea swells.

✳   ✳   ✳

Admiralty Bay is an oceanic refuge, a respite from the deep water, swift currents, and battering storms of Bransfield Strait. Its deep recesses are also a refuge from the packs of orcas that patrol the coasts where the penguins nest and the elephant seals haul out, hoping to find a careless chick or pup.[8] If the mysticete whales are the ecological and behavioral equivalents of the ungulate grazers on the East African savanna, the orcas are the hunting dogs, hyenas, and lions — the predators that hunt in packs. Once I walked along the beach at Potter Cove, an ice-free area just west of the mouth of Admiralty Bay, where the elephant seals were undergoing their annual molt, lounging like giant, mangy grubs on the beach. I watched a pod of five or six orcas swimming beyond the surf, occasionally peeking above the breakers at the beach. Theirs is a conformist society, and these orcas dived and breathed in near synchrony. When the wind luffed, their explosive breaths were audible well on shore. Now and then they bobbed their heads up into the air — a behavior known as spy-hopping — to peer at the seals on the beach. Orcas are creative hunters; they are known to bump ice floes with their heads in an effort to knock a seal or penguin into the water, and to toboggan on the surf in pursuit of seals. The sleeping seals hardly opened an eye for the orcas, but those caught in the surf undulated ashore as fast as they could.

Orcas are quintessentially social animals. Intelligent, highly cooperative, forming cohesive familial pods that bond for years (and probably decades), they are the antithesis of the socially amorphous baleen whales. The pods are led by a large patriarch. At 7,000 kilograms, he is about a third heavier than the adult females and is easily distinguished by his two-meter-long dorsal fin, which sometimes is so tall that it arches over onto itself. The rest of the pod consists of adult females and their calves, subadult males, and juveniles. The bull uses his mass to defend the pod against other bulls and to ensure his exclusive mating rights to the cows. This harem structure gives the bull a high degree of certainty that he has sired the calves in the pod, and socially this makes all the difference. In marked contrast to the indifferent paternity evinced by the male baleen whales, orca bulls take an interest in their young, gently playing with them, teaching them the ways of the sea, and sometimes overseeing them when the mother is away from the pod feeding.

Everything about the orcas implies social and cultural complexity, including sex. The orcas of the northeastern Pacific, which have been studied the most extensively, have at least four copulatory positions, a veritable *Kamasutra* by the unvarying standards of most animals. Even sex is cooperative; a third whale, usually a male, often holds the female in place during intercourse.[9] As with humans, sexual play and copulation are partly recreational. Juvenile males copulate with nonfertile females, perhaps as a means of reinforcing pod cohesion and of initiating neophytes into the behavior of adults. Although the act of intercourse is about as abbreviated as in the baleen whales, foreplay is prolonged and complex: the amorous male will typically slide in front of the head and dorsal fin of his fancy — a cetacean flirt — gently nudging her and forcing her to take notice.

As with the mysticetes, orcas maintain pod cohesion aurally, with a constant cacophony of whistles, squeaks, and chains of clicks. Orcas can probably recognize each other by voice alone, and pods may acquire their own dialects. But unlike the baleen whales, orcas and other odontocetes actively use sound to acoustically "visualize" their environment. They do this by producing high-frequency sound waves — animate sonar — and interpreting the angle, timing, and attenuation of the reflections. The toothed whales' concave skulls and bulbous heads, known as melons, are oil lenses, manipulated by muscles, that focus the sonar impulses. The reflected sound waves are gathered by a series of sinuses in the lower jaw.

The orcas are opportunistic flesh eaters. Very few animals have such a broad spectrum of prey species: fish, birds and, as a last resort, mammals, including baleen whales. Catching fish is probably a solitary endeavor, but it requires elaborate social organization for a pod of orcas, the largest of which weighs only 7,000 kilograms, to bring down a rorqual that weighs over 100,000 kilograms. Whalers have on occasion witnessed these attacks. One of the best descriptions was provided by Alan Villiers, who during the summer of 1923–24 worked as a crew member aboard the whaling vessel *Sir James Clark Ross* and watched a pod of orcas attack and kill a blue whale.

Two of the pugnacious killers attached themselves, one on either side, to the great mammal's lower jaw, and appeared to be bearing down on it with all the strength of their furiously struggling bodies, while two other of the assailants kept making short furious rushes at the great whale, and seemed to hurl themselves up and down on his exposed back, their writhing little bodies sounding on his body with loud thwacks. The fourth flitted between the point of the lower jaw, where he hung on with all his strength, and the broad back, where he joined the others in beating the life out of the poor whale. The harassed monster thrashed with his tail, beat his fins on the water with a terrible noise, and tossed his head from side to side, but all in vain. His spouts became feebler and scarcely rose three feet above his head. He was visibly tiring, and before long lay quietly on the surface, while the five killers turned their combined attentions to his lower jaw and dragged it open. Then as it hung down in the sea, they flitted in and out of the great mouth, gorging themselves on the tongue. They touched nothing else. They had attacked this giant for nothing more than the delicacy of the tongue.[10]

The toothed whales don't often visit the bay, at least not its recesses near Keller Peninsula, and the humpback calves are safe from orcas here. But beyond the mouth of the bay, anything goes. In the sea west of King George Island, I once sailed among a group of about thirty long-finned pilot whales, two minkes, and a solitary orca, all converging on a swarm of plankton. The mysticete whales were not afraid of the single orca; without the cooperation of his pack, he posed no threat. The sky above the whales was swarming with several thousand Antarctic and blue petrels, reeling and dipping into the sea. From a distance they looked like a cloud of gnats. It was a dark, snowy afternoon, overhung by glowering clouds that mimicked dusk and tricked the krill into coming close to the surface. The deep-water squids must have followed the krill to the surface, and life converged from below and above onto the thin border between air and sea.

On rare occasions, the seas around the South Shetlands are visited by sperm whales, the largest of the odontocetes. They are always solitary and always males, and unlike their smaller rela-

tives the orcas, they eat nothing larger than squid. The shape of
the sperm whale is unique in all of nature: a third of its length is
head and long thin jaw. Its blow also distinguishes it: always
angled forward and to the left.[11] Like orcas, sperm whales form
familial pods, but they are under the leadership not of a dominant
male but of a matriarch. The family pods remain in the subtropical and tropical seas year-round, where in the depths there are
plenty of squid. The adult males are either solitary or live in
bachelor pods and must joust with each other for periodic access
to females. Their tooth-studded lower jaws, rather than tools for
food gathering, are probably weapons that they turn on their own
species. Or on other enemies: there are numerous accounts of
harpooned sperm whales biting, and crushing, the small wooden
whaling boats that tormented them. This *machismo* has led to a
huge disproportion in size between the sexes. A socially mature
male may be as large as 64,000 kilograms, but a female will not
exceed 15,000 kilograms.

During the course of its life a sperm whale may be a member of
several social groups, each with its own purpose and geographical
range. After being carried by its mother for fifteen months, it is
born into a family pod consisting of other adult females and their
infants. The period of infant dependency and suckling is two
years, double that of the rorquals because of the need to learn
social skills. Socializing within the nursery pod continues until
the calf reaches puberty at nine or ten years. At that time, a female
calf may remain with the pod; she will briefly consort with the
visiting males and bear calves. But the male adolescent is driven
away (a mechanism to prevent inbreeding with its close relatives)
and joins a bachelor pod of sexually mature but socially immature males. There he may remain up to fifteen years before achiev-
ing the mass and social prowess to challenge the dominant males,
and gain access to females. Only then, and only on occasion, will
he consort with the females. Since he is never certain of the
paternity of the calves born into the pod, his parental interest is
negligible. Very large males spend much of their year in solitary

roaming. These are the sperm whales that journey to the Southern Ocean.

To capture squids, sperm whales dive prodigiously deep, to 1,000 meters or more. Marine biologists know almost nothing about the deep-water squids of the Antarctic, although so many species feed on them — the sperm, bottlenosed, long-finned pilot, and beaked whales, as well as the elephant seal — that they must be abundant. Squids are elusive, slippery, and athletic (they can swim slowly forward, explosively backward, and hover) and are difficult to catch, at least by humans, in the remote depths of the Southern Ocean. But the beaked whales, which have only two triangular teeth, and juvenile sperm whales, which have no teeth at all, just gums, manage to capture large quantities of deep-water squid. Over 18,000 squid beaks (equivalent to about ten days' feeding) were taken from the gut of a single sperm whale. Indeed, the best way to capture deep-sea squids for research often has been to use whales. Albert I, prince of Monaco, was one of the first marine biologists to realize this. In the 1890s he directed that his private yachts be used to capture sperm whales for the sole purpose of recovering the pens, beaks, and other body parts of giant squids in their stomachs. Even adult sperm whales that have been blinded (indeed, in the lightless depths greater than 300 meters, vision is useless anyway), or with partially missing jaws, have been captured with bellies full of whole, unchewed squid and no apparent malnutrition. How do they do it, blindly gumming their way through the deep, dark sea?

They may stun their prey with massive bursts of sound waves. Sound propagates easily in water, and a train of high-pressure sonic pulses, in theory at least, could immobilize squids and other prey long enough for them to be swallowed. How such a boom is generated in the melon of a whale is unclear; indeed how whales make *any* sound is not definitively known. But the melon of the sperm whale is the largest of all of the toothed whales, a complex structure of muscle, connective tissue, and spermaceti wax, and sounds are probably generated by compressing and rapidly squeezing air through the sinuses of the closed right respiratory tract. Sound therefore may not only locate the sperm whale's elusive prey but, once it is found, help catch it.

✳   ✳   ✳

In the long crepuscular hour I have lost sight of the humpbacks. Yet I can hear them breathing now and then when the wind is at peace. Bransfield Strait is losing its rage, and they will be gone in a few hours, at dawn. The terns are returning to the rock nests with one or two fish for their cryptic and hungry young. Ever afflicted with the constant worry of parenting, the terns scream at me as they fly by. They fly low over the bay, and across the beach, then, catching the updraft, climb the wind slope and dip into their secret swale.

11

The Passing
of the Leviathans

I must go down to the seas again, to the vagrant gypsy life,
To the gull's way and the whale's way where the wind's
like a whetted knife . . .

— JOHN MASEFIELD

THE SHORELINE of Admiralty Bay, and indeed many of
the beaches of the South Shetland Islands, is a charnel
house, heaped with whale bones looking like grotesque
sculptures. When the wind is easterly, the bergy bits jumble with
the bones on the beach. Beyond the Argentine station, on a bed
of green moss fed by the trickle of glacial meltwater, is the reas-
sembled skeleton of a rorqual. It was laid out during the summer
of 1972–73 by the crew of the *Calypso*, a distillation of the chaos
of bones scattered on the beach. Fifteen years later it still makes
good camera fodder for the red-coated tourists who trample the
beach and the moss every week. There are no conventional objects of tourist worship, no immutable monuments, no cathedrals
in this land of ever-shifting ice and cloud. And the works of
humans, except for a few historic huts, are depressing things,
ramshackle, intrusive stations and rusting oil drums. The whale's

spine is slightly curved, hinting of the arc of the smooth back in
the sea; its ribs are splayed on the moss, deflated. The smooth-
bowed jaws and skull are the biggest parts. The interior of the
vertebrae, where a fretting of bone once accommodated marrow,
blood, and nerve, is dusted green with algae.

✳   ✳   ✳

The first half of the nineteenth century was the heyday of whaling. In 1846 there were 966 whaling ships in the oceans of the
world, 736 of which were American. New Bedford, Massachu-
setts, was the center of the American whaling activity; at its peak
in 1857, that coastal town alone dispatched 329 whaling vessels
and 10,000 sailors. Until the middle of the century, whaling was
confined to the temperate and tropical seas and the Arctic Ocean,
which, surrounded by land masses, was relatively placid. Much
of the hunting concentrated on right whales, so named because
they were the "right" ones to seek: coastal animals, slow-moving,
docile when harpooned and, most important on the deep sea,
buoyant when dead. In addition to an abundance of oil from its
thick blubber, the right whale yielded high-quality whalebone,
which was used to make corsets and umbrella stays. After the
American and French revolutions, the pseudoclassical fashion of
feminine attire favored thin women. But after the coronation of
Queen Victoria in 1833, the caprices of courtly fashion shifted to
the ironclad, buxom femininity of stiff corsets and hooped skirts.
The right whale was doomed. Between 1804 and 1876, American
whalers killed 193,522 right whales in all of the oceans of the
world. The right whale was soon hunted to near extinction in the
North Atlantic, and the whalers turned their attention to another
coast-hugging species, the humpback whale, which in its turn
also became rare.

Having exterminated these easy prey, the whalers began world-
spanning voyages in pursuit of the sperm whale, which, although
lacking baleen, had a head full of waxy spermaceti, ideal for
making candles. And, like the right whale, the sperm whale did
not sink when dead. During the austral summer, when adult male

sperm whales migrated to cold polar waters, the tropical seas were filled with nursery pods of females and their young and with bachelor pods of socially immature males. These relatively small fry were easy prey for the whalers. Nursery pods consisting of mothers and their suckling infants were particularly easy to catch; once one of the whales was harpooned, the others would lend succor, often forming a circle with their heads pointed inward toward the injured whale, nudging it to the surface to breathe. But sometimes a large bull was hunted, and that could be dangerous. It was probably a solitary bull sperm whale that on November 20, 1820, attacked and sank the American whaler *Essex*, with a crew of twenty, in the middle of the Pacific Ocean, 3,200 kilometers off the coast of South America. The incident was the inspiration for Herman Melville's *Moby Dick*. Thomas Nickerson, a greenhorn sailor only seventeen years old, wrote of the attack,

> I then being at the helm and looking on the windward side of the ship, saw a very large whale approaching us. . . . I had scarcely time to obey the orders when I heard a loud cry from several voices at once that the whale was coming foul of the ship. Scarcely had the sound of their voices reached my ears when it was followed by a tremendous crack. The whale had struck the ship with his head directly under the larboard forchains at the water's edge with such force as to shock every man upon his feet. A second blow followed. The ship began to sink.[1]

*   *   *

Whaling in tropical and temperate seas during the nineteenth century was practical and profitable, but in the tempestuous Southern Ocean, in sailing barks at the mercy of the wind, using hand-held harpoons launched from fragile wooden catcher boats, it was much more difficult. Regardless, the allure of the south proved irresistible, and in 1849 the first attempt at whaling in the Southern Ocean was made by an Englishman, Charles Enderby, who with the support of James Ross started a whaling outpost in the remote Auckland Islands, south of New Zealand.

But in three years only one whale, a pregnant female, was caught there. "One morning a whale was reported," wrote R. E. Malone, an assistant surgeon at the colony,

> and the boats of the settlement started in chase; we followed in our slower boats to see the fun. After seeing him rise several times, the mate of one of the whale-boats struck him with a harpoon, fired from a harpoon pivot-gun in the bows of the boat. He rushed away with the line . . . till being well lanced and exhausted, he sank. . . . Two days later he rose. . . . One of the whale captains got in the chains, and, with a long-handled kind of sharp-edged spade, made the first, or, as they called it, the bride-cut; this he kept plunging in a zigzag way across the body behind the head, and then a parallel cut of the same kind near the tail, and then crossways. . . . On the first cut the escape of foul air was so great, and the stench so unbearable, that I was sick for some time, and the noise was like the rush of steam from the boiler of a steam-engine when the steam is eased off. . . . It was a young cow-whale come to calve, about 45 feet long, and yielded five tuns of oil and some whalebone. This was the first and only whale ever taken in the islands, notwithstanding all the expectations and the numbers seen in the bay; and it was quite a holiday in the settlement, every one, even the women, some of them with children at the breast, came to look at the beast.[2]

So the Enderby whalers left the sheltered bays and turned to the capricious ocean. Their effort was dangerous and futile. In 1852 the whaler *Hardwicke*, after four months at sea, during which time one sailor died, four were placed in the brig, and most of the crew contracted scurvy, returned to Enderby Island with hardly any whale oil. The colony, and any hope of whaling in the Southern Ocean, was abandoned.

But Enderby and Ross's dreams would not fade. In 1874, in the Scottish whaling town of Peterhead, the brothers David and John Gray published a modest booklet, *Report on New Whaling Grounds in the Southern Seas,* a reiteration of the observations of

Ross and his crew, thirty-three years earlier. During the next two decades the booklet was reprinted several times in Scotland and Australia, and it proved influential in whetting the appetites of whalers at a time when northern stocks, particularly around Europe, were becoming depleted. As a result of this momentum, the Antarctic Exploration Committee of the Royal Geographical Society was created. The committee sent four vessels to take stock of the whales in the Weddell Sea during the austral summer of 1892–93. They were the celebrated Scottish "Dundee fleet," consisting of the *Balaena,* the *Active,* the *Diana,* and the *Polar Star.* The Royal Geographical Society contributed £150 for the purchase of scientific equipment for the ships, but the purpose of the expedition was commercial. The same year, Norwegian entrepreneurs independently organized another expedition to the Weddell Sea, on the bark *Jason.*

The Dundee expedition soon realized that Ross's descriptions of the abundance of right whales had become obsolete; very few right whales were left in Antarctic waters. During the fifty years since Ross sailed in the Weddell Sea, the right whales had largely been killed off, not in their Antarctic summer feeding grounds, but in the fjords and bays of Patagonia where they spent the winter. If there was to be whaling in the Southern Ocean, it would have to seek the abundant humpbacks and the swift blue, fin, and sei whales. At the time of the Dundee expedition, whalebone was selling for £3,000 a ton, and the yield from only three or four fin whales would pay for one trip.

But the big rorquals were formidable prey. In desperation for a quick profit, the crew of one of the wooden whaleboats from the *Active* managed to sink three harpoons into a fin whale, which swam off like a juggernaut, boat in tow. Soon the harpooner of a second whaleboat embedded three more irons in the whale, and a sailor from a third boat managed to grab the streaming lines and join the fray. When the whale surfaced to breathe, rockets were fired into its flanks. But it was all to no avail. For fourteen hours the untiring whale hauled the three boats through the cold sea, until at last it snapped the lines and swam off with six harpoons embedded in its flesh and trailing several hundred yards of two-inch rope.

Surrounded by whales but unable to catch them, the expedition turned to seals. "It is useless to trust to the sealers for exploring purposes," complained C. W. Donald, the scientist aboard the *Active,* "for as long as they can fill their ships with blubber in latitude 64° they will never penetrate to 65°."[3] The *Balaena,* 260 tons and equipped with a 65-horsepower steam engine, was the largest and best equipped of the Dundee fleet. She was under the command of Captain Alexander Fairweather, a myopic despot who disdained scientific inquiry and sought only profit. Except for a brief encounter with the aptly named Danger Islets in the Weddell Sea, she was never closer than nine kilometers to land. "A man who would work a hired dog . . . as these poor beggars are worked every day ought to go and hang himself," wrote W. G. Burn Murdoch, the expedition's artist and chronicler. "A dog's food is sumptuous compared to their tasteless, monotonous diet." The slaughter of seals, considered by the captain to be the only labor of any value, went on seven days a week, and landings were forbidden for any purpose but sealing. "Of all the hypocritical, canting humbugs in the wide world the Lowland Scotch sectarian is out and away the worst," griped Murdoch.

> The bigotry without the justice of his covenanting forebears has surely come down to hum wi' muckle aggrandisement. Would any one who knows what a Sunday in Dundee means, believe that a crew from that godly, radical town would be ordered to put aside the laws of God to work at sealing on the Sabbath day? Yet so it is, and we are told that for this Old Testament law-breaking we have the sanction of the worthy Presbyterian minister . . . and the crew who have been killing seals all week, day and night, are sent off dog-tired to paint the snow vermillion.

Murdoch, the first person to record the vistas of these new (and frustratingly distant) southern lands, was forbidden to work on

Sunday and in desperation had to retreat to his cabin and clandes-
tinely sketch from memory.

> Though sealing is "a wark o' nacesseety," bawbees being gained
> thereby, "drawing pucturrs on the Lord's Day o' Rest is an awfie-
> like thing!" This I am told very clearly and explicitly, and have to
> do reverence to the Creator's works by painting from memory in
> the privacy of my bunk . . . cursing at the same time the length of
> my legs and the interference with the purest form of worship.[4]

By January 29 the decks of the *Balaena* were heaped with the
skins and blubber of 4,800 seals. Wrote Murdoch, in obvious
frustration over the lost opportunities:

> Would that I owned this ship and this good crew even for three
> summer months in the Antarctic. . . . One vessel, or two in con-
> sort, could chart the whole of the unknown southern continent.
> Think of this, ye rich, who dream of knighthood and more riches!
> For £10,000 this chance is going, cheap I call it — a chance to
> write your names in Big Type on the maps of the world. Think of
> this, ye gentlemen of England, who yacht at Cowes in ease, the
> chance is going — going; and if you don't bid for the South Pole,
> some bold Yankee and his fair lady will be down there before
> you get under way, and then — there will be no new place under
> the sun![5]

The *Jason* was under the command of Captain Carl Anton
Larsen, an urbane and curious Norwegian who had already spent
four seasons sealing in the Arctic and who, as we shall see, was
destined to transform the Southern Ocean. In contrast to the
Dundee fleet, conditions aboard the *Jason* were clean and the
food was good. On January 21, 1893, Larsen entertained the captain and disillusioned officers of the *Balaena* aboard the *Jason*.
As Murdoch described it,

> Captain Larsen received us on the poop with a Norwegian wel-
> come, and we got down to the cabin to the tune of "Way down
> the Swanee River," played on an organ on deck. They really do
> things well on a Norwegian whaler, and it seemed by the variety

of provision below as if the *Jason* had just left port the day before. There was milk just like fresh milk, rusks, and liqueurs, and all sorts of good things; and such coffee — nothing to speak of in Norway, but what a contrast to our noxious mixture on the *Balaena!*[6]

That season, Larsen explored the coast as far south as 65°. While the Dundee fleet stayed far to sea, he landed on Seymour Island, where he recovered fossil leaves and wood from an eroding bank. "When we were a quarter of a Norwegian mile from shore," he wrote in his journal, "and stood about 300 feet above the sea, the petrified wood became more and more frequent, and we took several specimens, which looked as if they were of deciduous trees; the bark and branches, as also the year-rings, were seen in the logs, which lay slantingly on the soil." These were the first fossils recovered in the Antarctic since Eights's discovery in the South Shetlands, sixty-four years before. But the economics of slaughter tugged even at Larsen. "How interesting it would have been to explore that land!" he wrote, "but, as we are not sent out for scientific exploration, but for whale and seal hunting, we had to resist the temptation."[7]

In January 1902 Larsen returned to the Weddell Sea as captain of the sail- and steam-powered sealer *Antarctic,* on an expedition that was to prove pivotal in initiating Antarctic whaling. It was led by the Swedish geologist Dr. Otto Nordenskjöld, thirty-two years old, who planned to overwinter with six other men, while Larsen and the *Antarctic* returned to the Falkland Islands. As he sailed among the South Shetland Islands, encountering the inevitable cycle of storm and tranquility, Nordenskjöld wrote in his journal, almost prophetically, "The weather had changed as if by magic; it seemed as though the Antarctic world had repented of the inhospitable way in which it had received us the previous day, or, maybe, it merely wished to entice us deeper into its interior in order to more surely annihilate us."[8] After exploring the Pacific coast of the Antarctic Peninsula, Nordenskjöld entered the Atlantic at Erebus and Terror Gulf, at the northeastern tip of the peninsula. He chose Snow Hill Island as his land base for over-wintering. The island fascinated Nordenskjöld because, like nearby Seymour Island, it had an abundance of fossils. Among other notable discoveries during the long winter on Snow Hill Island, Nordenskjöld found the remains of an extinct giant penguin.[9]

In the meantime Larsen and the *Antarctic* had returned to the Falklands to spend the winter. However, the following spring, when the *Antarctic* attempted to return to Erebus and Terror Gulf, the weather fulfilled Nordenskjöld's prophecy, and Larsen found his ship locked in pack ice. His first mate, Gunnar Anderssen, set off on foot over the sea ice in an unsuccessful attempt to bring the Nordenskjöld party to the open water in Hope Bay, at the tip of the Antarctic Peninsula. Larsen continued his attempts to nudge the *Antarctic* into the pack ice, but even using both steam and sail, he made little progress. Finally, eighty kilometers northeast of the Antarctic Peninsula in the Weddell Sea, the *Antarctic* was crushed and sank. Larsen and his crew, pulling open boats laden with supplies across sixty-four kilometers of pack and polynya, struggled to Paulet Island, arriving on February 28. Immediately they prepared to spend the winter. They constructed a hut of stones, roofed with canvas and insulated with sealskins. The walls were only a meter high, but in the center of the hut, supported by tent poles, a man could stand upright. The chinks between the stones were caulked with penguin guano. The work, hastened by autumn storms, took only a week. Paulet Island has one of the largest Adélie penguin rookeries in western Antarctica, and during the summer the flanks of its hills are crowded with penguins. Before winter set in, Larsen and his crew killed 1,100 penguins and stacked them, frozen, near the hut as provisions for the winter.

Nordenskjöld and his crew prepared to endure another winter on Snow Hill Island, and Anderssen and his crew dug in at Hope Bay, where they constructed their own stone hut and where there was also a bountiful supply of Adélie penguins. Each of the three parties overwintered unaware of the fate of the others. The following spring Larsen and his men, and, in a separate party, Anderssen and his crew, walked across the frozen Erebus and Terror Gulf to Snow Hill Island, where they were reunited with Nordenskjöld and his companions. They must have been quite a sight, encrusted with the filth of two Antarctic winters. By this time both Argentina and the European nations were concerned about the fate of the expedition, and in the early spring of 1903 the Argentine vessel *Uruguay,* reinforced to withstand the ice, rescued all parties.

The ordeal was more than an epic of exploration and survival in the Antarctic. It was the great crucible for Carl Anton Larsen, who within a year returned to the Antarctic to begin large-scale whaling. After his rescue, at a banquet in his honor in Buenos Aires, Larsen made a fateful speech, recorded in his accented English. "I thank jouse vary mooch, and dees is all vary nice and jouse vary kind to mes, bot I ask jouse ven I am heer vy don't jouse tak dese vales at jour doors — dems vary big vales and I seem dem in houndreds and tousends."[10]

Some in the audience, regaled by this consummate Antarctic survivor, decided to invest in Antarctic whaling and founded La Compañía Argentina de Pesca, S.A. Larsen was given a license by Britain to base a factory ship equipped with boilers to render the blubber, and a single catcher boat, at Grytviken in South Georgia's sheltered Cumberland Bay, where he installed himself as manager of a crew of Norwegians.[11] In mid-December the first whale was shot in Cumberland Bay; that austral summer 194 more whales, mostly humpbacks, were killed. The following year, while Larsen was in Norway buying provisions, he rallied other whalers to venture to Antarctica. During the summer of 1905–6, C. C. Christensen, a Norwegian whaling entrepreneur infected by Larsen's enthusiasm, sailed the *Admiralen,* a factory ship, to New Island in the Falklands and later to Admiralty Bay.

During this period the bones began accumulating on the margins of the bay. In these early days of Antarctic whaling, only the blubber, rich in oil used as fuel for lamps and as a lubricant for machinery for the burgeoning Industrial Revolution, was retained. The rest of the carcass was discarded and, bloated with the buoyant gases of decomposition, floated to the nearest beach. Thomas Bagshawe, who as a youthful adventurer journeyed to the South Shetland Islands during those early years of whaling, described the scene at Foster Harbor, on Deception Island, which must have been identical to what was occurring in Admiralty Bay:

> [The vessels] in the harbour . . . only extracted the oil. Consequently when these ships had finished with their whales, the carcasses and other refuse were allowed to float around them. The stench from the masses of intestines, stomachs and livers was too revolting to be described, and the water around the ships was discoloured with the oil and blood. The shore in the neighbourhood was hidden by an accumulation of meat and bones in various stages of putrefaction. Some of the flesh was a relic from previous seasons, so that, when getting ashore from a boat, one was liable to find oneself wallowing in old whale flesh instead of stepping upon solid ground.[12]

During the summer of 1906–7, the first Chilean factory ship, the *Gobernador Bories,* belonging to the Sociedad Ballenera de Magallanes, began operating in the South Shetland Islands. Argentine and British whalers soon followed. By the 1912–13 whaling season, twenty-one floating factory ships, six shore stations, and sixty-two catchers were operating in the South Shetlands, the South Orkneys, and South Georgia Island, and the catch for that season was 10,760 whales. Profits increased, and so did the carnage. As one species of whale became scarce, the whalers turned to the next. Since the right whales were already practically extinct, the whalers turned to the docile humpbacks, which stayed in coastal waters, making it easy for the catchers to haul the carcasses to the factory ships anchored in the sheltered bays. But within a few seasons the humpbacks, too, were rare, and like the Auckland Island and Dundee whalers before them, Larsen and his comrades had to face the daunting rorquals.

But by the turn of the century, technology had caught up with the whalers' ambitions. The developments were largely the work of another energetic Norwegian, Sven Foyn, who beginning in 1864, when he was already wealthy from sealing, developed the modern whaling technique, variations of which were used until the 1980s. Foyn invested $20,000, a fortune in his time, in perfecting his method, which used fast, steam-powered catcher boats, equipped with a bow-mounted cannon that fired a harpoon armed with hinged barbs and a twelve-and-a-half kilogram cast iron grenade. The detonation of the grenade inside the body of the whale often killed it instantly (although some whales required several shots), and the barbed harpoon and restraining line prevented the animal from sinking. The straining of the whale and the buffeting of the sea were absorbed by a series of springs, called the accumulator. Once the whale was winched next to the ship, it was buoyed by compressed air injected through a pneumatic hypodermic lance. The heroic days of whaling were forever over. Soon after whaling began in the Antarctic, it became a mechanized, industrial grind, an insatiable maw consuming the once invulnerable rorquals. At first it was most economical for the industrial whalers to seek the largest prey, the blue and fin whales, because the expense of detonating a harpoon in a large rorqual was the same as in a small one, and the explosion was usually just as lethal. They disdained the small sei and minke whales as not worth the effort or the expense of a harpoon.

And so, in succession, the blue and fin whales became rare in the waters surrounding the Antarctic Peninsula and South Georgia. The catchers had to journey farther out to sea and tow the carcasses farther back to the factory ships. As the stocks of whales declined, the government of the Falkland Islands Dependencies belatedly became concerned and, for the season of 1912–13, set a limit on further expansion of whaling in what it considered its territorial waters. Had this measure been followed, the stocks of Antarctic whales might not have been decimated. But again technology kept pace, and the pelagic factory ship, which could operate on the high seas, was developed, again by the Norwegians. Its invention was a matter of accident. The summer of 1912–13 was unusually severe, and the pack ice around the South Orkney Islands did not break up until late spring, preventing the Norwegian factory ship *Thule* from entering its usual safe harbor in the islands. Out of necessity, the *Thule* was tied to the edge of the pack and the whales were processed alongside. This momentous development freed whaling from that time on to take place beyond the territorial waters, and control, of any nation.

In 1923 Antarctic whaling spread to the Ross Sea, and the dream of Ross and Enderby at last came to fruition. The pelagic factory ship *Sir James Clark Ross,* a converted merchant vessel, and five small catchers battered their way through the pack ice that guards the entrance to the Ross Sea and began a relentless carnage. Larsen, at age sixty, was in command, having left retirement on his farm in Norway. On the day after Christmas 1923, the first whale, a large male blue, was killed in the Ross Sea. Alan Villiers described the kill:

> Despite the explosion in his body of the soft iron shell with which the harpoon was tipped, the great monster refused to die. As long as he could he remained below, struggling madly to free himself, but at length he was forced back to the surface again for air. As he rose the powerful winch on *Star II* [a catcher] quickly hove in the harpoon line, and as his great back broke the surface of the sea the gun was loaded again and Captain Iversen, standing like a waiting matador, was ready to administer the death thrust. As the sorely wounded whale lay wallowing and struggling in bloodstained foam the captain carefully trained his gun at the heaving target, and just as the bull turned in a vain effort to dive again, lifting half of his great bulk out of the water, the gun spoke once more and a second harpoon flew into the mountain of blubber and flesh. Instantly he sounded, the harpoon line running furiously over the bow-wheel. But not for long. The second shot had told, and scarcely two minutes later the great whale lay, a mountain of death, suspended on the harpoon line fathoms below the gently rippling surface. And thus the first whale died.[13]

During that season in the Ross Sea, the *Sir James Clark Ross* killed the largest blue whale ever measured, a female 31.4 meters long. She was perhaps the largest animal on Earth at the time and bigger than any dinosaur, one of the largest animals ever to have existed in the long history of life on the planet.

In spite of the carnage, the expedition proved unprofitable. Although the ship had patented German meat extractors, much of the carcass, as in the South Shetlands two decades before, was discarded. But Larsen, not a man to give up, returned to the Ross Sea the following year. He died on that voyage, peacefully, in his bunk. He was stern and frugal to the end. "If that old fellow up there [Larsen] sees you throwing [even] half a shovelful of coal into the sea he comes rushing down here like a madman," a boatswain aboard the *Sir James Clark Ross* admonished Villiers. " 'We want all this coal,' he says. Why, he goes around the deck picking up old nails . . . although what use he could put the nails to he did not explain. . . . He knows what it is to have no coal, no food, and no nails."[14]

✳   ✳   ✳

Typical of Antarctic whaling in the 1920s were the factory ships *Gouvenor* and *Telefon* and their satellite fleets of catchers, which were based in Admiralty Bay. The catchers sailed into Bransfield Strait to kill whales and tow them back to the bay, like wolves returning with food to their den. The ships were not equipped for polar seas. The *Telefon* was sunk by the pack ice at the mouth of the bay, at a place now memorialized as the Telefon Rocks.

The catchers were designed for the hunt: 15 to 20 meters long, low-beamed, and swift, able to chase whales at speeds up to fifteen knots. All hatches and ports had to be sealed, even in moderate seas, and unless the crew had a reason to work on the lurching and slippery deck, they were confined below. Their quarters were cramped and without privacy. Unlike the sealing vessels of a century before, the whaling ships had steam heat and hot water, and the men were clean and warm. But in the damp warmth of the cramped quarters, dripping with mildewed clothing and permeated with the reek of the bilge, the stench must have been just as nauseating.

The crew of a catcher consisted of the captain, a gunner, mate, chief and second engineers, and several stokers and ordinary seamen. The gunners were pivotal to the whole enterprise and were one of the first professional groups to organize a union in Norway. A skilled gunner earned bonuses for everybody aboard a factory ship; an unskilled one, singularly responsible for the poor catch of a season, won the derision of everybody aboard. Four-hour watches were posted, with shorter "dogwatches" from four to six and from six to eight in the evening. "Blaast! Blaast!" would shout the watch from the crow's nest. Once a whale was sighted, it would be chased, often for several hours, until it was within range and proper angle of the gun. There were no heroics here, no Nantucket bravery. Wrote Villiers,

> With a spurt of flame, a deafening crash and roar, a flying of twisting, writhing wads through the air, a pungent, burning taste of gunpowder, the gun speaks and the ponderous two-meter length of forged steel is expelled, making the heavy rope from the bow-wheel tray fly through the air in all manner of fantastic loops and figures. The leviathan has been hit and sounds instantly in a blinding fury of foam and spray. The water before the ship is a seething cauldron. A deep red tint is seen. A dull thump — the explosion of the shell deep in the whale's vitals — comes from the depths.[15]

The main harpoon lines, seventeen centimeters in circumference and 220 meters long, were stored in two lockers on either side of the bow. Up to four of them could be spliced together into a line that was nearly a kilometer long. The lines were fed onto the deck over a winch drum, which served as a brake, and were suspended from the mast by a pulley-block, itself anchored below deck to the massive springs of the accumulator. A thinner, lighter fore-line was attached to the harpoon and coiled on deck next to the cannon.

The preferred spot to shoot a whale was in the middle of the body beneath the dorsal fin; this destroyed the heart and lungs. After firing, one crew member minded the winch drum to ensure that first the fore-line, then the main lines, did not play out too fast, throwing sawdust or coal dust on the rope to decrease friction. Another crewman reloaded the cannon, this time with line from the second locker.

Frantic, the harpooned whale would gradually play out the line, sounding and then rising for breath, its innards exploded, exhaling blood, vulnerable once again to the cannon. Some harpooned whales would tow the 150-ton catcher bucking through the sea, until slowly, yard by yard, the whale was winched closer and a second shot could be fired. The sailors marked the distress of the whale by the interval between blows; a strong whale, in spite of the fatigue of the chase, would breathe every minute or so, then, as it weakened, every twenty seconds, . . . then ten. At last, mortally exhausted, the whale exhaled and sank. The corpse was winched to the surface, bobbing in its own blood and a sheet of oil, with Dominican gulls and giant petrels turning in the chaos of moving sea and clattering equipment. Killer whales and leopard seals would sometimes arrive on the scene and take bites from the tethered whale. A sailor leaped onto the carcass and thrust the inflating tool into the corpse to pump it with air, a temporary measure until the gases of decomposition bloated the corpse.

The corpses were winched to the side of the catcher, flukes toward the bow, head astern. The fore-line was severed and the flukes were trimmed, for in rough weather their slapping would dent the iron hull of the ship. Notches were carved in the stumps of the flukes according to the number of harpoons embedded in the whale's flesh, so that the workers aboard the factory ship would not blunt their flensing knives on embedded metal or accidentally detonate an unexploded charge. "The massive harpoons are often twisted into weird shapes by their contact with the whales," wrote Bagshawe, and each catcher had a blacksmith, whose job was to reforge them.

Life aboard the factory ships was a toilsome monotony of cold, stench, and gore. "There were two kinds of days — bad days, and worse — and each lasted twenty-four hours," wrote Villiers.[16] The whales that had been dead the longest were the first to be processed. Their insulating blubber prevented the freezing ocean from chilling the meat and viscera; "any that have been lying for some time have swollen to a terrific size and look as if a mere pinprick would cause an explosion," wrote Bagshawe.[17] In the early days, before stern-loading factory ships arrived in the Antarctic, flensing — the stripping of blubber from the corpse — was started as the whale floated in the sea next to the ship. Two flensers, standing in a tippy flat-bottomed boat or on the slippery body of the whale itself, lacerated the blubber with broad, long-handled knives, threaded a wire rope through the skin, and winched the blankets of fat aboard the deck of the ship. "[The flensers'] hands are bare," wrote Villiers. "They can not wear even fingerless mitts, for they must have sure and steady hold of the greasy knives — one false sweep, one wild cut, might spell destruction for themselves or their mates. Frequently they must cease work for a moment, plunge the knife into the hot flesh, bathe their hands in the warm blood to bring life back to them."[18]

Stripped of its blubber, the corpse was towed to the opposite side of the factory ship, where it was dismembered. "Great joints are cut from the carcase," wrote Bagshawe. "The flensers wallow up to their waists in the body whilst they hack away with their sharp knives. It made me feel violently sick to see the casual way they cut off and chewed quids of tobacco regardless of the blood and mess on their hands."[19]

On deck the crew continued cutting up the carcass like an army of ants swarming over a fallen bird. The blubber was further cut into chunks, passed through a chopping machine, and cooked in huge steam-driven meat and blubber boilers. "A boy works with each cutter," wrote Villiers, "holding the blubber steady with a steel hook. Both the blubber-cutter's and the boy's faces are covered with blood, which freezes quickly and sets hard upon them, giving them a most piratical look."[20] The whale's tongue, the item so prized by marauding orcas, was also especially sought by the whalers. Intricately banded and cross-banded with muscles, and therefore "wonderfully difficult to walk on," in the words of one flenser, one blue whale tongue could yield up to 2,000 kilograms of oil.[21] Wrote Bagshawe:

> As work on board got into full swing, the ship looked a disgusting site and had a still more disgusting smell. The decks were soon covered with a mixture of whale oil, blood, and coal dust. To pass forward there were two alternatives, either to go by way of the top deck and wade through slush with the risk of slipping down in it all, or to go by way of the lower deck with drops of blood and fat splashing upon you through the planks above. The first journey I made was by the latter route, foolishly enough without a hat. I thought that the moisture dripping on to my head was water until I found out, to my horror, that it was whale blood. The smell of it all was awful.[22]

The residue from the boilers was shoveled into the sea, to add to the heap of slowly rotting, frozen sludge that accumulated around the ship. "Woe betide any poor unfortunate who happens to be on the rope-ladder behind the waste pipe as it escapes," wrote Bagshawe, "for he is soon enveloped in a foul cloud of the most pestilential odour imaginable."[23] Indeed, the reek seems to the pervading motif of the chroniclers of Antarctic whaling. As Villiers wrote,

> Even those who have been working among whales for years have to become accustomed afresh every voyage to the stench, and as for the greenhorns, we were sick for a week. The smell was everywhere after that first whale was caught. It was in our food, in our coffee, in the water with which we washed, in our clothes and in our bunks. We were steeped in it.[24]

✳   ✳   ✳

While the packs of factory ships and their catchers scourged the South Shetland Islands, the South Orkneys, and the Ross Sea, shore-based whaling stations were constructed in Foster Harbor on Deception Island and in the sheltered fjords on the eastern, leeward side of South Georgia on the same beaches that had been stripped of fur seals a century and a half before. On South Georgia the stations were given names from the North Sea, a half-world away: Stromness, Leith, Husvik, Grytviken. Although licensed by the British, the stations were mostly managed and manned by Norwegians. These were the first settlements in the Antarctic, complete with houses (for the managing elite), hospitals, theaters, and social halls. There were even a few women and children, but only the managers were afforded the luxury of female companionship. The workers suffered as wretched an existence as their seafaring brethren.

A saying among whalers at the time was "beyond 40° is no law; beyond 50° no God." To remedy this situation, Captain Larsen personally financed the building of a church at Grytviken in 1913. High-steepled, with white wooden walls and a tin roof, it was built behind the barracks and bath house. At the time it was the southernmost church in the world.[25] When the Grytviken church was consecrated on Christmas Day, the clergyman described his congregation as "young men in the prime of life, all weather-beaten and hardy, clearly bearing the marks of their toil," and lamented that "Christian life unfortunately does not wax strong among the whalers."[26] The church had no permanent pastor, only a succession of clergymen, and in 1931 it was converted to a community hall and warehouse. Over the years a graveyard sprouted around the church, wooden monuments to men lost at sea or mashed by machinery. Despair and loneliness in the ladyless company of whalers, frequently resulted in men being sent home as mentally deranged. A station manager at South Georgia reported that one whaler asked whether he could spare a new piece of rope; he wanted to hang himself, but the only rope he could find was rotten.

They were in the south only for the high wages. Whalers in the Antarctic earned two to three times what they could earn at a shore whaling station in Norway. "I think it's strange that the manager on a floating factory and the mates pay no attention to the scenery around them, they're only looking for whale," wrote a visitor. "They don't see the beauty of an iceberg, they don't look at the birds or anything else of interest."[27] During their free time the whalers on South Georgia indulged in hobbies peculiar to their trade. One collected whale eyeballs, which he hollowed out, dried and formed into parchment lamp shades. Another fashioned the skin of the heart into lady's handbags and tobacco pouches. Others carved whale bone or scrimshaw.

In March 1913 the workers at Leith Harbour struck, protesting poor food, barracks infested with vermin, and the station doctor, whom they labeled "a Scottish quack." A magistrate summoned from Grytviken promptly arrested eighteen striking whalers and deported them to Norway. But by 1920 the situation had considerably worsened. According to the Colonial Report for the Falkland Islands Dependency, thirty-six "Russian Bolsheviks" fomented a strike at one South Georgia station, intending to establish the first Marxist republic outside of the Soviet Union, and rallying "all workers in all whaling stations all over the world to organise."[28] The effort collapsed when, by chance, a British warship on routine patrol, its captain and crew unknowing of the insurrection, entered the harbor. Outgunned, the Bolsheviks surrendered.

The whaling stations on South Georgia are all derelict now; the last was abandoned in 1964. The buildings are ramshackle and cluttered with debris, spattered by squalls of snow and rain. Things move strangely in the wind, as if haunted; broken walls batter their frames, rusty cables sway, doors lurch on their rusty hinges. The hymnals, blotched and moldy, are still in the pews of Larsen's church, but the pews and altar are caked with rat feces. Tourist ships visit the stations every summer, and the subdued visitors wander through the wreckage with a peculiar reverence, as if visiting the mute remains of an ancient civilization. They are admonished by their guides not to take souvenirs. But in fact these ghost settlements are themselves vandalous acts on a pristine landscape, graffiti left by an industry of such rapacity that it extirpated its prey and extinguished its own fires.

✳   ✳   ✳

During the First World War, when humans turned on themselves the same destructive energies that they had once directed at the whales, the demand for Antarctic whales increased. It was the first "modern" war, and the large-scale use of artillery bombs whetted the world demand for glycerin, then derived primarily from whale oil, to make explosives. The pressure on Antarctic whales increased for political reasons as well. Britain threatened not to renew the Norwegian whaling licenses in her Antarctic territories, so Norway, utilizing the experience gained from the *Thule,* built its first pelagic factory ship with a stern slipway, the *Lancing.* Whaling on the high seas, independent of shore stations and regulations, became common.

After the war the slaughter of whales by pelagic fleets continued unbridled. During the summer of 1930–31 alone, 40,201 whales were destroyed in the Southern Ocean, surfeiting the European market with whale oil and causing prices to drop. The widespread electrification of Europe and North America, along with the mining of petroleum, had diminished the demand for whale oil used in lighting. But technology, particularly the newly invented process of hydrogenation of liquid oil to make fats for soap and margarine, again kept pace, opening new markets for whale products. The manufacture of margarine, which sold for one-fifth the price of butter, was a self-defeating industry for Britain and Norway, since both nations had huge domestic dairy industries. But there were markets abroad for fats, particularly in impoverished and war-devastated Germany. By the 1930s the German dependence on imported fats had strangled its domestic dairy industry and was depleting foreign currency reserves.

Adolf Hitler's efforts to prepare for another war were being re-
tarded. In 1933, declaring a "war on margarine" and turning
butter to guns, Hitler set quotas on the importation of fats.
Germany by then was outfitting its own whaling fleets and sending them into the Southern Ocean.

Across the world, Hitler's allies the Japanese, were also looking
to the Antarctic. For centuries the Japanese had been hunting
whales in their coastal waters and eating whale meat. But during
the 1930s an increasingly industrialized and expansionist Japan
began to seek more far-flung hunting grounds. The Japanese
began pelagic whaling by emulation, first by sending workers to
the whaling stations of Norway and later by chartering entire
fleets of factory ships and catchers, complete with crews, from
the Norwegians.

By 1939, thirty whaling fleets from five countries — Norway,
Britain, the United States, Germany, and Japan — were scouring
the Antarctic. But during the Second World War, whaling in the
Antarctic practically ceased. By then glycerin could be synthe-
sized from petroleum, and all men, materials, and ships were
diverted to the conflict. Factory ships were converted to cargo
carriers, and many were sunk. Before the war there were forty-
one floating factory ships; after the war only nine remained. Yet,
in unpredictable ways, war once again sealed the fate of the
Antarctic whales. One invention that emerged from the war was
to have far-reaching effects on whales: the jet turbine. The high
pressures and temperatures of turbines required specialized lubri-
cants that were not easily distilled from petroleum. But sperm
whale oil — the spermaceti that was made into such everyday
household items as candles during the last century — was ideal
for this purpose.

After Japan's defeat, General Douglas MacArthur authorized
the resumption of Japanese whaling as a temporary measure, "an
emergency humanitarian move," under the control of occupation
authorities. In 1946 Japanese pelagic whaling resumed in the
Antarctic, and during the early years of the American occupation

47 percent of the meat consumed in Japan was from whales. (That same year Norway loaned the Soviet Union a fleet of whaling ships.) But the Allies' concern for the welfare of the war-displaced poor was not entirely humanitarian. One of the conditions under which Japanese whaling was permitted was that the oil would be transferred, often on the high seas, to North American and European vessels. The Allies also tried to flog their whale meat within their own borders. In 1946 "whale burger steak," part of a consignment of ten tons of frozen whale meat from the British whaler *Empire Victory,* was served at a luncheon at the Institute of Refrigeration in London. A newspaper reported that "those who ate it claimed that it even tasted like chopped steak." It didn't catch on with Europeans and North Americans, but the Japanese, with their long tradition of eating whale, always provided an eager market. Even so, as better substitutes developed for most whale products, whaling became an orphaned industry. More and more, whales were used to make nonessential items, such as pet food, cosmetics, and soap.

The whaling fleets of the 1950s and 1960s were floating armadas, pelagic industrial sites into whose stern slipways were winched most of the remaining Antarctic blue and fin whales, and many of the sei and minke whales. Typical of the new ships was the British *Balaena,* named after the Dundee sealer that first prospected for whales in the Weddell Sea sixty years before. At 15,000 tons, she was equipped to process 1,500 whales in a season and store their meat in a quick-freezing plant. She was accompanied by a fleet of ten catchers armed with high-voltage, electrocuting harpoons, whale-seeking radar, and two amphibious planes. Only 80 members of the crew of 550 were British; the rest were Norwegians.

The whales didn't stand a chance, in spite of the formation in 1949 of the International Whaling Commission, a generally impotent body with no authority to enforce its recommendations until it was too late. To its credit, the I.W.C. banned the hunting of right, humpback, and blue whales — but only after their populations had shrunk to the point that it was no longer economical to pursue them. The I.W.C. set whaling quotas for its member nations, but the quotas were defined not in terms of the species or their rarity, but in terms of "blue whale units," an ineffective measure that continued until 1972. One B.W.U. equaled two fin whales, two and one half humpbacks, six sei, or twelve minkes. The absurdity of the B.W.U. is illustrated by the fact that it remained the standard even after blue whales had become so rare that their hunting was banned.

Today only vestigial numbers of right, humpback, blue, and fin whales remain in the Southern Ocean. In 1989 the Scientific Committee of the I.W.C., using new data based on direct sightings, estimated that several hundred right whales remain in the Southern Ocean today. There are only 3,000 to 4,000 of the 100,000 humpbacks that originally lived in the Southern Hemisphere; about 500 of the 180,000 Antarctic blue whales; 2,000 of the 400,000 Antarctic fin whales; and 1,500 of the 150,000 sei whales. No Antarctic blue whale has been killed in the Southern Ocean since 1962, but there is some question as to whether enough blue whales remain for mates to find each other on the lonely seas and to rebuild the population.

Not all rorquals declined in numbers, however. The little minke whale, which because of its small size was not hunted except as a last resort, may have actually increased its population, largely because of the surplus of krill left by the extermination of its larger competitors. Although the minke has only small amounts of blubber, it does yield good-quality meat, and even though killing minkes was unprofitable for the Europeans, it remained economical for the Japanese.

One by one, during the 1960s and 1970s, the European whaling nations pulled out of Antarctica. Antarctic whaling seemed to be declining, but the Europeans merely sold their fleets to the Japanese, who continued whaling, unabated, in the same vessels. Finally, the I.W.C. declared a moratorium on all "commercial" whaling in the Antarctic starting with the 1985–86 season. Japan eventually complied, but only after taking 1,941 whales during the following summer. However, the ban did not preclude the taking of whales for "research" purposes, in order to determine sex ratios, migration patterns, pregnancy rates, and the age composition of the whale populations.[29] This information had been repeatedly collected over the decades, and showed that the population structure of the whales has been shifting in a manner that reveals clues to long-term changes in the Southern Ocean. Because the food supply for the minkes has been plentiful, they have been growing faster over the last few decades, and they have begun to reproduce at an earlier age (from fourteen years in the 1940s to only six years in the 1970s). These changes profoundly affect the reproductive rate of the whales and the entire food balance of the Southern Ocean.

The hunt for whales continues, in spite of international protests, particularly from the United States, which in the 1960s abandoned whaling and adopted the self-righteous purity of the newly reformed. During the summer of 1987–88 Japan killed 300 minkes in the Southern Ocean, and although the declared reason for the slaughter was "research purposes," their scientific motivations didn't prevent them from selling the meat. And on November 10, 1989, the Japanese factory ship *Nisshin Maru No. 3*, departed from Japan for the Southern Ocean; in December she was joined by three catchers. The plan was to catch another 300 minkes.

The Soviets also continued pelagic whaling into the 1980s. Unlike the Japanese, their prey was in large part sperm whales, and their objective may have been to secure a supply of turbine oil to keep their huge fleet of military and civilian aircraft in flight. By the 1970s Americans had discovered that the oil extracted from the jojoba bean, a native of the American Southwest that had been long used for cosmetics and soaps, was an ideal substitute for spermaceti in jet turbines. But a guaranteed supply of jojoba oil was at first unavailable to the Soviets. Later, the plant was cultivated in the arid zones of central Asia, and there were no longer any strategic reasons for them to continue whaling. It was the final unpredictable development: the sperm whales of the world were saved by a desert shrub.

It may at first seem difficult to understand why the whaling nations drove their quarry to the verge of "economic extinction," the level below which they are so scarce that it is not profitable to continue hunting them. It would appear to be suicidal behavior for any enterprise. However, a closer examination of the economy of whaling reveals a tragic imperative. The immediate profit from the unchecked destruction of whales, when invested in the economies of industrialized nations, ultimately yielded more income than their sustained harvest would have. Many of Norway's port facilities and much of her modern-day merchant fleet were built with the profits from whaling. Ultimately and tragically, conservation may never have been in the best economic interest of any of the whaling nations.

*　　*　　*

In the twilight I walk, beneath the cloud-dark, to Cousteau's whale skeleton, near where I first landed in Antarctica six years before. The moonlight, scattered by the clouds, pervades the night sky. A clutter of icebergs, snapping loudly in melt, is scraping the pebbly beach. The glacier is bright on its edges, as if it has been buffed. The terns are roosting and are at last silent. It is hard to imagine the bay during its time of carnage. I try to imagine these fogs that envelop the moon as greasy smoke, soot, coal dust. Instead of silence and the tinkling of melting ice, the squeal of winches and reverberation of steam engines. The bay sheened with oil. Instead of this clarid, cold air, the stench of rendered flesh and decomposition. This still beach, littered with whale-rocks, heaped with rotting guts, undulating in the surge.

The skulls and vertebrae that litter the beach, just fragments now, are many times larger than I. The imps who killed these whales are long gone, but the skulls remain, sixty years after their death, hard as stone, resistant to ice and cold, dark winters. The wages of the whalers have been drunk or gambled away. Only the white and green bone-rocks remain.

I walk across the moss beds to the whale skeleton, sit on its wide and comfortable cranium, walk down its ribs, up again to the huge arched jaws, then sit once again on the occiput. In the darkness, blustered by gusts of cold black wind, I can almost feel the animal surge in the swells. Did this whale, seventy years ago, swim through the stormy Bransfield Strait, feeding on the ancestors, twenty, forty, sixty generations ago, of the krill that swarm to my light? Was it a male? Did it sing its beaconing chant into the dark, deep sea and, in the autumn, follow the nubile females to the tropics? Was it a female? Did she briefly consort with a male on the high seas, and did a calf grow in the warm shelter of those lower ribs, twice carried across the Drake Passage? Being a composite, it did all of these things. This whale is Everywhale.

12
The Tempest

O, for a draught of vintage! that hath been
Cooled a long age in the deep-delved earth,
Tasting of Flora and the country-green . . .
— JOHN KEATS

YESTERDAY WE WERE RADIOED the details of our return
to the north. The *Barão de Teffé* will carry us thirty-two
kilometers to Chile's Teniente Marsh Station, and from
there a Força Aérea Brasileira Lockheed C-130 will fly us to Punta
Arenas in Chilean Patagonia. But the *Barão* is out of radio contact, and the schedule is uncertain. An antsiness has set in. It is
late in the season, and the days of calm weather are getting rarer.
There was a hard snow last night; winter is close.

Now the manic weather has changed once more, and today
there is a warm tempest: the mercury edges up to 8° C. The wind
slings rain at the windows, swells the lichens on the mountain
slope, and tousles the bay. I watch the needle on the wind gauge
in the meteorology module; the gusts are in excess of 100 kilometers per hour. I am rushing to finish the last experiments —
photomicroscopy of the parasites I have studied. These may be
my last days in Antarctica, my last chance for discovery. But the
wind vibrates the laboratory module so violently that images
bounce in the field of view, and I have to wait for the gusts to pass
before pushing the shutter.

Forms were passed out this morning: psychological screenings, recommendation forms for those who want to overwinter — what claustrophilic person would want to do this, in absolute isolation? We prepare *relatórios,* descriptions of our activities for the bureaucrats and military comandantes in Brasília. A summer's work is condensed into a few pages: here is my sliver of a contribution to Antarctic science. In the lounge, everyone is partying. Our collective attention span has snapped. Since I started writing this page, the music blasting from the stereo has changed twice, and now *Marathon Man* is on the tube. The din lasts until three A.M. Thank heavens for earplugs. I sleep fitfully, and too much.

At dawn I walk, leaning into the wind, beyond the abandoned Argentine station, to take a last look at the whale skeleton. The beach is in the wind-shadow of Keller Peninsula, and the sea is smooth. But high above, the wind shears vapor off the surrounding glaciers, the clouds dissipating into sky, and across the bay I can hear booming surf and groaning icebergs. Here, in the lee of the storm, a much colder, moister air mass rolls off Stenhouse Glacier and oozes over the beach. No doubt it bears the eroded bubbles of ancient atmospheres. It is a deadening wind that insinuates itself into my parka, my gloves, my hood. The erotic summer of life and light has ended.

At twilight the mountains behind Arcktowski are silhouetted against a tarnished horizon. The storm has brought an ice fog to the glaciers, giving them false height and smoothness. The entire island is generating ice fogs, becoming enshrouded like a lost world. On the steep, tumbled slope behind the beach, streams of glacial melt are chortling. But as I listen in the night, the air temperature passes that pivotal point of freezing, and the streams stop their earth-sounds. Three gentoo penguins stand as still as ice on the crepuscular beach. I sit nearby, invisible in the dusk. After a long moment one of the penguins, evidently experiencing a slight hormonal surge — a vestige of the departed season of chicks and sociability — squawks. Suddenly, all three waddle to the sea and jump in. I can see the pulses of their wing beats bulging the metallic sea, as rapid as a stone skipping across the water, and, far away now, a white triangular tail. The impending winter has brought this sea change. The summer is forgotten; they are aquatic now.

The tide is lapping high on the rocks, and I walk on the soft moss-beach, covered with crisp snow, melted and congealed again. I almost step on a sleeping Weddell seal, born this season, lying prostrate on the beach. It ignores me. It breathes intermittently with explosive breaths, as if it were diving, flexes its foot-fins, shivers its whiskers, and distorts its face: rapid eye movements. Of what is this baby dreaming? Its quickly departed mother? Delicious slow-moving fish? Leopard seals under the ice floes?

There is a rush of wings in the steeping dusk. A skua has come to investigate me. Is it asserting its bleak little territory; has it come to see if I am dead and edible? In the hills behind the beach at this midnight hour, his family complains, not the usual large vocabulary of cooperative hunters, but a plaintive, lonely hawk-cry.

✳   ✳   ✳

For three days the storm lays siege. And then at dawn it is calm. The *Barão de Teffé* anchors in a sullen sea above my dive site. Taking advantage of the calm, we load the ship in haste. Pickled fish and amphipods, microscope slides. My data book and journal, sealed in Ziploc bags, never leave my hands. The Zodiacs drop the twelve crew members who will overwinter, and the *Barão*'s two helicopters fly in their equipment and deposit it on the beach. Hundreds of crates: corned beef, frozen chickens, coffee, *cachaça,* tinned peaches, mattresses, magazines, paper . . . They have to think of everything; the *Barão* won't return for nine months.

By noon we have hoisted anchor and are sailing past the fossil forest and sealers' camp at Cape Hennequin, the disconsolate Madonna at Arcktowski, past the abandoned penguin rookery at Point Thomas and into the bouncing Bransfield Strait.

Antarctica, so removed from the mainstream of human life, haunts one's memories like no other place I know, and already I am nostalgic.

These are the last wild days on Admiralty Bay. The advance of bickering humans, with their redundant research stations, their imposition of national frontiers on the southern continent, their garbage dumps, air strips, and septic pits, seems inexorable. The bay barely survived the onslaught of the sealers and then the floating factories of the whalers. Both industries invaded Antarctica when it was profitable to do so, and both left when the profits ceased. Now the resources of Antarctica beckon profit once more. There is krill to feed the ever-gaping maw of humanity to the north. There may be oil and natural gas. Antarctica can no longer have an icy indifference to humans.

Even if none of this avarice descends on Admiralty Bay, there will inevitably be global environmental changes, the result of industrial activities in the north. Carbon dioxide levels are rising. The whole planet, but particularly its polar areas, may be warming because of the greenhouse effect. Since 1950 there appears to have been a slight but significant increase in summer mean temperatures in the Antarctic. Antarctica, at least along its margins, may be slowly thawing. In Admiralty Bay the glaciers are retreating farther into the cold fastness at the center of the island, and the Ternyck Needle stands higher and more barren than ever. On Heard Island, where once there were only eleven species of vascular plants, three new species have appeared during the past few years. On Signy Island, the retreating glaciers have exposed beds of viable peat between 2,000 and 3,000 years old. Experiments have been conducted on that island in which transparent shelters were placed over barren soil, creating, in effect, miniature greenhouses. Within fifteen years, up to eight species of mosses and ten species of lichens appeared in the greenhouses. Antarctic winters in the future will still be long and frozen by temperate standards. But could warming due to carbon dioxide create a similar greening throughout coastal Antarctica?

The greenhouse also affects the sea. With the thermal expansion of water, sea levels may be rising in Admiralty Bay, indeed all over the world. Someday the black rock beaches of Keller Peninsula may disappear under the ice-littered sea. Cousteau's whale skeleton may at last be buried at sea, and Stenhouse Glacier, calving ever faster now, will coat the new ocean floor with silt, creating new habitat for mud-dwelling amphipods and *Serolis*.

In recent years ozone, a particularly reactive form of oxygen, has become depleted over Antarctica during the early springtime. Its loss is due to complex reactions with catalytic chlorofluorocarbons, artificial compounds used in, among other things, refrigeration, the electronics industry, and the production of foam plastics. The reactions take place in the stratified winter clouds, unstirred by the warm breath of convection. If all of the ozone that is dispersed in the Earth's upper atmosphere were concentrated into a single, pure layer, it would only be about three millimeters thick, yet this vanishingly small quantity of gas filters mutagenic ultraviolet light from the sun, including the particularly penetrating ultraviolet B, 290 to 320 nanometers in wavelength. The effects of the increased UV flux on the productivity of the photosynthetic lichens and mosses, and on the plankton of the Southern Ocean, are unknown. Some species of phytoplankton, equipped with melanin and other sun screens, are already protected from ultraviolet light. Others will inevitably decrease in numbers. Even if the overall productivity of the Southern Ocean does not diminish, the *quality* of the productivity may change. For example, there may be a shift to diatom species with thicker silicon shells that are indigestible to krill. And when krill don't eat, everybody starves.

The greenhouse effect and ozone depletion are the great inadvertent experiments of our time. Both will have a particularly severe, although not necessarily harmful, effect on Antarctica. Using models drawn by supercomputers, we soothsay the future, but only our descendants will ever know for sure. Yet Antarctica may ultimately offer a germ of hope for our species. When twelve nations signed the Antarctic Treaty on December 1, 1959, they forged the first multinational arms control treaty to be negotiated since the end of the Second World War.[1] The Antarctic Treaty covered all ocean and land south of 60°, equivalent to about 10 percent of Earth's surface. The seven signatory nations that claimed Antarctic territories — France, Norway, New Zealand, Australia, Great Britain, Argentina, and Chile — all placed their claims (which covered about 85 percent of Antarctica) in abeyance and, in an early display of solidarity, the United States and the USSR agreed to neither establish territories of their own nor recognize the claims of others. Regardless, scientific stations represent a de facto claim — in effect squatters' rights — to Antarctic territory. And the American base at the South Pole represents a tacit occupation of all of the wedge-shaped territorial claims (except Norway's) that intersect there.

By 1991, thirty-nine nations had signed the Antarctic Treaty. However, it is an uneasy alliance. Only twenty-five countries have been given full "consultative" status, meaning they are allowed to participate in the meetings that determine the fate of Antarctica, conduct inspections of other bases, and modify the treaty.[2] According to treaty rules, consultative membership requires a demonstrable research presence in Antarctica. This explains the plethora of new "research" stations in the remaining ice-free areas, built by nations, such as the People's Republic of China and Peru, that can little afford an Antarctic extravagance — an ironic by-product of the treaty that has caused Antarctica more harm than protection. Even Greenpeace, in an effort to become a consultative party to the treaty, established a permanent base of its own, on Ross Island next to Robert Falcon Scott's hut.[3]

Because Antarctica is a vast, uninhabited wilderness that some nations might consider ideal for nuclear tests, the treaty specifically prohibits the deployment and detonation of nuclear devices as well as the disposal of radioactive waste. In effect, Antarctica was designated a nuclear weapons–free zone — the first ever negotiated by humans.

These are wonderful events, with few precedents in human history. Through the Antarctic Treaty, our species — at least on paper — has thrown down its arms and disdained territoriality on the last earthly continent. But the treaty is mere paper, and it has prevented neither belligerence nor weapons of war from reaching Antarctica. Some would argue, for example, that the Falkland Islands war, fought between Argentina and Britain in 1982, was as much a conflict over favorable strata for natural gas in the islands and the nearby Antarctic Peninsula as it was for nationalistic honor.[4] Ironically, it was garbage that provided the initial cover for the war: an Argentinian scrap merchant, contracted to dismantle the derelict whaling stations on South Georgia, surveyed the island in preparation for the invasion.

And the treaty has not prevented us from bringing our newfangled nightmare toys to Antarctica. During the austral summer of 1961–62, a 413-ton "portable" nuclear reactor, affectionately dubbed "Nukey Poo" by base personnel, was installed by the U.S. Navy at the American base at McMurdo Sound. The reactor was placed on the slope of Observation Hill, a promontory that had been used as a lookout during the expeditions of Robert Falcon Scott and Ernest Shackleton early in the century. The facility was greeted with the optimistic hubris of the second decade of the atomic age. Admiral George Dufek, the former commander of the American military task force in Antarctica, wrote that Nukey Poo "marks a revolutionary step in polar exploration. . . . It opens a dramatic new era in man's conquest of the remotest continent, and incidentally fulfills an old dream of mine."[5] The reactor was said to be a cheaper and less polluting alternative to the diesel generators then in use at McMurdo. The plant ran intermittently for eleven years, but in a hostile environment where energy ensures survival, it was never reliable. The cooling system of the plant continuously released radioactive argon-41 into the atmosphere, and on three occasions cracks were discovered in the containment vessels, leaking water into the soil. Finally, because of a mechanical fault, Nukey Poo was permanently shut down in

September 1972, and McMurdo was converted back to diesel power. The following summer was exceptionally cold, and the shipping lanes into McMurdo Sound were blocked by ice until early autumn. Diesel fuel was in such short supply that the base was nearly evacuated and closed.

Nukey Poo was cut up and hauled away. The U.S. Navy insisted that no high-level radioactive waste remained on Observation Hill after its disassembly. "You would have to lie on the dirt for three months to acquire as much dose as a chest X-ray," said a navy spokesman at the time. Yet to comply with the terms of the Antarctic Treaty, the navy had to remove 12,200 tons of radioactively contaminated rock from the site and ship it to the United States to be buried. Now a piece of one of the historic landmarks of the age of south polar exploration has wound up, at great expense, in a toxic landfill half a world away.

Seven years after the demise of Nukey Poo, on the other side of the continent, the Antarctic environment experienced its second, and much more ominous, nuclear event. On September 22, 1979, an American Vela surveillance satellite, designed to detect atmospheric nuclear tests, recorded the bright binary flash characteristic of a nuclear explosion in the vicinity of Prince Edward Island, a remote subantarctic outcrop claimed by the Republic of South Africa. A few hours later, astronomers at the giant radio antenna at Arecibo, Puerto Rico, detected a ripple of radio waves in the ionosphere that could have been a sequela of the blast. The observations were soon shrouded in ambiguity, contradictions, and secrecy. A panel of scientists convened by the Carter administration concluded in 1980 that the "event in the South Atlantic" was "technically indeterminate," ruling out cosmic rays, solar flares, sun glint, lightning bolts, even "super lightning bolts," and was due to an event for which there was no explanation (known in the lingo of surveillance as a "zoo animal"), such as sunlight reflecting from the facets of a turning meteor. But a report issued the same year by the U.S. Navy Research Laboratory concluded that the evidence supported a "nuclear event." A CBS News crew

reported that the flash was from a secret Israeli nuclear test conducted in cooperation with South Africa. Other sources pointed
the finger at the Soviets. In the babel of confusing claims, the
issue remained unresolved until 1991, when the government of
South Africa finally acknowledged the blast. Humans, it seems,
have brought their most primitive, and most modern, tools here:
stone arrow heads and nuclear weapons.

✳   ✳   ✳

The pilot of the Força Aérea Brasileira C-130 radioes that he has
left Punta Arenas. He is bringing on this flight to King George
Island supplies and a delegation of Brazilian generals, admirals,
politicians, and reporters. The plane is to arrive in only two
hours; the absolute fastness, the isolation of Antarctica no longer
seem real. We are taken by Zodiac from the *Barão de Teffé* across
Maxwell Bay, past Great Wall Station, its sentinel cat in the
window, through a banner of oil excreted by Bellingshausen Station, to the dock at La Villa de las Estrellas. Several construction
workers are indolently sitting on a barge, eating their lunch.
There is a queue at the bank; one needs cash to get by here. The
grocery store has a new shipment of fresh Chilean peaches and a
few cabbages. Squealing children, swaddled in thick parkas, kick
a soccer ball on the gravel road. They are beautiful children, red-
cheeked and happy. To them, the comforts of this place must be
indistinguishable from those of the suburbs of Santiago. Declining a ride in a truck, I decide to walk the two kilometers to the
Teniente Marsh air base, past the ranch houses, the green Russian
meteorological station, the heaps of rocks and gravel, the runway
lights that have been switched on in anticipation of the flight.
Everything is bulldozed, scraped away. This station has become a
massive scar. At the top of the hill in front of the hotel, a group of
workers is setting up a new satellite dish made of prefabricated
aluminum. Just beyond, on a pavement of tarmac, two helicopters and a red Twin Otter are being preened by mechanics.

The FAB plane has arrived. It is camouflaged like a hunting
bird: pale tawny underside, green on top. There is no starter for a

C-130 at Teniente Marsh, so it keeps one engine running. The delegation steps from the airplane into the warm, thawing wind and bright sunshine, muffled in thick down parkas fringed with wolf fur, padded mittens, and cumbersome Sorel boots. They plant their first footsteps in Antarctica on the tarmac in front of the hotel. A television reporter and a cameraman emerge from the plane, blinking at the brightness, and hastily set up shop; they will be in Antarctica only two hours. The camera starts rolling and, like a self-declared Columbus, the reporter, lost for words, states the obvious. "This is Antarctica!" he blurts, sweeping his arm toward the horizon, careful not to pan the camera toward the runway and buildings. A few Chilean military officers look on with dyspeptic unease; the indifferent workers keep hammering and sawing in front of the hotel — perhaps they can get the antenna ready in time for tonight's Chilean soap operas.

The delegation will be taken by helicopter to Comandante Ferraz, will pose for photos, have a toast or two at the bar, cancel some postcards, and return in less than an hour. In the meantime, as the helicopters clatter off, I sip sodas in the hotel with scientists from the nearby Korean station. The airplane is being loaded with cargo through its wide rear door. In exchange for using the airstrip, FAB has agreed to transport a load of Chilean cargo to Punta Arenas. It is tons of garbage, the refuse of the good life at Teniente Marsh, stacked in pallets and covered with nylon webbing. Not much else is exported from King George Island these days. The garbage is beginning to thaw on this warm afternoon and is dripping. Is this a statement by the Chileans, a graceless reminder of their sovereignty? Or is this just good housekeeping? The load master weighs every pallet, every scrap, and positions it carefully in the hold; an imbalance could cause the plane to careen out of control.

The Brazilian delegation is back now, with stacks of postcards. Some people are clutching rocks with branching *Usnea* lichens, doomed to dry out as desktop ornaments in the tropical halls of bureaucracy. I explain to a general that he is removing an organism that may be hundreds of years old, older even than the nation of Brazil, older than the European arrival in the New World. "Puxa!" he shrugs. "Who could imagine such a thing?" But he keeps the rock. What's the use of dropping it on the tarmac?

We climb into the dark belly of the plane, stepping over the miscellany of expedition gear: duffel bags, aluminum boxes, skis, and straps. All seem to have been piled randomly into the high interior of the plane, but in fact everything has been meticulously balanced by the load master. The walls are covered with green padding and drip pulleys, clips and hydraulic tubing; the floor is cold aluminum with sockets for the cargo pallets. The plane is already beginning to smell of garbage. There are only six windows on a C-130, the minimum necessary to monitor the condition of the wings and engines, and there is a scramble to sit near them. I preempt a window next to the door, forcing all subsequent passengers to stumble over my legs in the darkness. There are no true seats on the airplane, just loosely woven red nylon straps slung over an aluminum framework. The straps seem spaced precisely to cut off the circulation in one's thighs. The door is sealed. I sit with my back to the window, contemplating the dark load of cargo. I study the label on an empty green wine bottle, Gato Negro, and the pretty maritime scene on a tin that once held mussels. A crew member, his work finished, has already dozed off on top of a pallet of garbage.

The undulating gravel runway at Teniente Marsh ends abruptly atop the cliff face on the north of the island, launching the planes over the freezing sea, where there are no boats, no hope of rescue. But the C-130 is ideal for these conditions. It is comfortably powered by four big turboprops, and its high wings diminish the chances that the engines will inhale the stones spit up by the wheels. The FAB pilots are as good as any I know. But in the gloomy and lurching body of the plane, listening to the changing pitch of the props, I wish that it weren't burdened with so many tons of garbage.

The plane takes most of the runway, lifts just as the cliff face drops into the sea, and then is grabbed and tormented by the gusts spinning off the Southern Ocean. The damned, incessant wind hounds us even as we leave. The garbage shudders and the nets restraining it squeak and stretch. Somebody is throwing up, and the acid-sweet stench slides through the hold. The general sitting next to me tries to appear coolly indifferent, contemplating his fingernails, clutching his ancient, lichen-encrusted rock.

✳   ✳   ✳

Punta Arenas, Chile. Everything is linear in this city: roads, rows of telephone lines, the blocky poured-cement buildings. Punta Arenas is the southernmost city in the world, a settlement that grew in defiance of the persistent Patagonian wind, which sculpts and frazzles the vegetation, bending the beeches and junipers and causing them to seemingly cringe in agony. There are fences against the wind: tightly fitted slats or hedgerows of junipers. The buildings have a thick-walled stolidity, and when one steps inside, there is that stuffy, prickly sensation of total insulation. No leaky windows let in fresh breaths of air.

The pier at Punta Arenas, jutting into the Strait of Magellan, is the embarkation point for the Antarctic expeditions of a half-dozen nations. It is the last contact with the thermal world. The *Polar Duke,* a research vessel used by the U.S. Antarctic Program, the *Barão de Teffé,* the *Cruz de Froward,* the Chilean navy, the tourist ships, all replenish here. But tonight at the pier there is only a five-tiered sheep barge, reeking of straw and manure, loading thousands of disconsolate animals for the slaughterhouses of Valparaíso, and a tourist ship. The tourists stop at the Club de la Unión and dine in the vine-tangled greenhouse, which presents long windows to the warm north. But the scientists and crews prefer Sotito's Bar or the Hotel Ritz, near the base of the dock and across the street from the pillared gymnasium, which is a replica of the Parthenon. Once these were scruffy bars, full of seafaring folk, where one could buy a good beer and some *mariscos,* but now they have been discovered and have wine cellars and use cotton tablecloths. The waiters evaluate you with a critical eye.

Punta Arenas is enveloped in a thin sea fog. The strait at eleven P.M. looks like heat-tempered blue steel. On the other side a small fishing village twinkles under a brooding hill. It is tranquil tonight, but this is a sign that tomorrow may be a screamer.

The next morning there is a cyclone, born in Antarctica, and on the hilltops it is hard to stand against the wind. On the strait the wind actually lifts sheets of water and slings them against the shore. Planes of clouds, each moving in a different direction, laminate the sky. A few of my Brazilian colleagues and I drive 380 kilometers across Chilean Patagonia, first along the margin of the sea, then over rolling tundra to the fishing village of Puerto Natales.

Some aspects of Patagonia are nearly oceanic: treeless expanses of sheep-cropped grass and cushion plants. One can watch the wind eddies from kilometers away. Occasionally we see a lonely homestead huddled in the expanse of plain and sky, surrounded by a dense green windbreak of junipers. Saline lakes splotch the countryside, their rims hoary with salt. Some have flamingos, puffy pink in the wind; all have ducks. It is late summer, the end of the dry season, and the savanna between the trees is burnt umber, spattered with yellow dandelions and daisies pushing their heads above the sheep-chewed grass. In the swales of this hilly land, where there is protection from the full force of the wind, are the southern beeches. Most grow in small groves. Their crowns, just hinting of orange autumn colors, are bearded with pale green *Usnea* and spherical parasitic mistletoes.

We are sensually delirious. Antarctica has become a reverie. Antarctica, ultimately, is an inarticulate land. And upon one's return to the north, even to the wind-blasted islands of Patagonia, it seems as if the senses have come alive again. One is reacquainted with the smell of life: of decomposition, pollen, a blossom. How eloquent the north is! Even food tastes better.

There is a modest restaurant in Puerto Natales, La Ultima Esperanza (The Last Hope) that may serve the finest salmon on Earth: thick, pink, workingman's salmon steaks, grilled slightly smoky and served with bright, fresh vegetables and dark wine. The fjord in front of the Coffee House of Last Hope is adrift with wild birds — zither-calling gallinules, whistle-voiced oystercatchers, and mute black-necked swans. Dolphin gulls, with alarmingly orange bills and legs, stand sentinel on the dock awaiting the return of the fishing fleet.

We camp in a sheltering copse of beeches at the base of the Torres del Paine, a range of sawtoothed mountains fronted by fjords and long freshwater lakes. Not far away is a herd of guanacos, New World relatives of camels. Guanacos have big, moist eyes, like children in the cheap portraits painted on velvet and sold in dime stores. The females and infants have a soft, playful friskiness, and the males are randy and combative. They bleat threateningly at each other and, in a tangle of slender leg and neck, joust on the slopes while the females demurely watch. The wind is also mad, disheveling the landscape and nudging the tussock grass. Only the obdurate beeches refuse to swoon to the wind's bluster.

It is bitingly cold, but too windy to light a fire. We retreat to our polar sleeping bags, spreading them on soft grass slightly scented with the barnyard smell of guanacos. This place whispers of Antarctica, and in a way we have traveled back in time since leaving Admiralty Bay. Was Admiralty Bay like this 16 million years ago, when groves of southern beeches cloaked its hills, before Antarctica froze? At dawn I awoke, steeped in the warmth of my sleeping bag, now rimed by frozen dew that has swept down the mountain in huge breaths during the night. Each beech in the little copse where I am camped has a wind-voice of its own: shudders, sighs, whispers, and long teeth-chattering groans. The beards of *Usnea* drip disconsolately. The tundra and mountains are possessed by sky. Two Andean condors, soaring higher, ever higher, on gusts and thermals, are a double helix in the wind-blue sky. Now the white-headed male tilts on broad wing and slips; his mate, in graceful dissymmetry, follows.

Afterword

I T HAS BEEN TWELVE YEARS since I flew from King George Island in the reeking FAB C-130. I have not returned to the Crystal Desert. I am a professor now, a vassal to schedules and deadlines. I no longer have the luxury of an extended, unfettered period of time at the end of the year. This is fine with me. The world is a splendid orb — much too diverse to be understood in a single lifetime — and I've become absorbed in other wonderful places. I'm sure I will return to Antarctica someday when duties decline, and she will greet me like an old friend.

Yet just as the ancient Greeks assumed that there had to be a vast southern continent — a *Terra Australis Incognita* — to counterbalance the northern land masses, my memories of Antarctica temper and balance my perspective on the rest of the planet. Those memories have taught me what Earth was like before it was mantled in life, and how she would appear if we humans were to run amok. Antarctica's erose coastline, where the ebullient cold sea blooms around the clock and, for a season, flourishes, taught me of life's voluptuous capacity to prosper just about anywhere. On this winter solstice — the shortest day of the year in the Northern Hemisphere — I watch a murder of crows turning darkly over my snowbound garden in Iowa, gobbling the last morsels of a dead squirrel before the early onset of night, and I know that at this moment there is an antipodal place of perpetual sunshine, where the skuas never sleep and the penguins are ever vigilant. Last summer, as I was strolling along the Bund of Shanghai amid a sea of people, I remembered a place where there are no people. Since leaving Antarctica, I've dived in the tepid, hungry waters of Rio Negro and remembered a place where the water is so cold that the fish have antifreeze in their blood. I've trekked in the shaggy forests of Belize, where whole Maya cities are buried, and remembered a continent where the most ancient human relic is less than two centuries old. I have camped in the western Amazon when it has been rocked by *friagems*, the vernal storms that shatter the forest, creating the mosaic of gap and canopy — and, therefore, the forest's diversity — and understood that those winds were spun off the angry circumpolar ocean only a few days before.

I have found congruencies, too. Since leaving Antarctica I have journeyed by icebreaker to the North Pole and noted the strange similarities between Arctic and Antarctic: the tartar of sympagic algae beneath the white pack, the abounding snow-draped mountains, the robin's-egg blue of a glacial till–tinted fjord — a pastel extraction of a terrain smothered by ice. These convergences teach us that there may be rules of assembly — rules of order — on Planet Vita.

The Arctic tern greeted me in both places.

And I remember the singularities. The heavenly light, for example. I've never lived in a place where light mattered more than in Antarctica, where the photons seemed newly minted. Since leaving there I've become a connoisseur of light, always dissatisfied in northern latitudes, where the air bears the detritus of erosion and the clutter of life: Sahelean dust, spider silk, and spore.

Today more tourists than scientists visit Antarctica: 12,000 of them sailed there in the year 2000. Every one, I'm sure, was making a pilgrimage and returned home transformed. This is as it should be. Antarctica should no longer be the exclusive province of government-funded scientists, whalers, and krillers. It should become a world park — the first ever — accessible to all people but possessed by none.

During the past twelve years King George Island has become the Hamptons of Antarctica, burdened now with the "research" stations of ten nations: Argentina, Brazil, Chile, the Czech Republic, Germany, the People's Republic of China, Poland, Russia, South Korea and Uruguay, all crowded together on the shoreline. Collectively, those nations fought eight wars among themselves during the twentieth century. Now, at the bottom of the world, they share a sure camaraderie, placing their territorial claims in abeyance and learning to live peacefully and cooperatively as neighbors. The Antarctic Treaty was designed, at least in part, to show us a way out of the cold war, and it seems to have done that. On a continent where the blood of no mother's son has been spilled in defense of homeland and where no brother has slain brother, people have laid down the tooth-and-claw lessons of savanna and forest, of city-state and nation-state and, by necessity, learned tolerance.

This may be Antarctica's greatest gift to the rest of the world. Is it the way of the future? Perhaps not. But for a while, at one place and for one brief time in human history, we are living up to our name: *Homo sapiens.*

Grinnell, Iowa
December 21, 2001

Appendix

Latin Names of Plants and Animals Mentioned in Text

PROTISTS

Diatoms (Phylum Bacillariophyta)
    Opalescent diatom   *Thalassiosira* spp.
Green algae (Phylum Chlorophyta)
    Penguin guano algae   *Praseola* sp.
    Red snow algae   *Chlamydomonas* spp.
Red algae (Rhodophyta)
    Calcareous algae   *Lithothamnion* spp.

FUNGI

Yeasts (Class Ascomycetes)
    Amphipod parasite   *Candida frigida*

LICHENS

Bipolar lichen   *Cladonia rangifera*
Coastal crustose lichen   *Buellia latemarginata*
Orange crustose lichen   *Xanthoria* sp.

Yellow crustose lichen — *Caloplaca* sp.
4,000-year old crustose lichen — *Rhizocarpon geographicum*
Marine lichen — *Verrucaria serpuloides*
Fruticose lichens — *Usnea antarctica*
Usnea fasciata
Usnea sulfurea
Himantormia lugubris
Umbilicaria sp.

VASCULAR PLANTS

Mosses (Division Bryophyta)
  Bipolar moss — *Polytrichium alpestre*
Conifers (Division Coniferophyta)
  Araucarias (Family Araucariaceae)
    Admiralty Bay fossil (extinct) — *Araucaria* sp.
    Mt. Augusta fossil tree (extinct) — *Araucarioxylon* sp.
Flowering plants (Division Anthophyta)
  Cabbages (Family Cruciferae)
    Kerguelen cabbage — *Pringlea antiscorbutica*
  Guaraná Family (Sapindaceae)
    Guaraná — *Paullinia cupana*
  Grasses (Family Gramineae)
    Bundle grass — *Deschampsia antarctica*
    Tussock grass — *Poa flabellata*
  Pinks (Family Caryophyllaceae)
    Antarctic pink — *Colobanthus quitensis*
  Orchids (Family Orchidaceae)
    Macquarie Island orchid — *Corybas macranthus*
  Beeches (Family Fagaceae)
    Antarctic southern beech (extinct) — *Nothofagus* sp.
    Patagonian southern beeches — *Nothofagus* spp.
  Beans (Family Leguminosae)
    Jojoba bean — *Simmondsia californica*

ANIMALS

Sea stars and sea urchins (Phylum Echinodermata)
   Brittle stars (Class Ophiuroidea)   *Ophionotus victoriae*
   Sea urchins (Class Echinoidea)   *Abatus shackeltoni*
Spiny-headed "worms" (Phylum Acanthocephala)
   Amphipod parasite   *Corynosoma* sp.
Sponges (Phylum Porifera)
   Glass sponges   Class Hexactinellida
Tardigrades (Phylum Tardigrada)   *Macrobiotus furciger*
Arthropods (Phylum Arthropoda)
   Spiders and mites (Class Arachnida)
      Mites (order Acarina)
         Free-living mites (Suborder Trombidiformes)
            Red mite   probably *Stereotydeus palpalis*
         Orobatid mite (Suborder Sarcoptiformes)
                             Alaskozetes antarctica
         Soft-bodied mites (Suborder Actinedida)
            Nunatak mite (5° from South Pole)
                             Nanorchestes antarcticus
   Springtails (Class Collembola)
      Moss-dwelling springtail   probably *Cryptopygus antarcticus*
   Insects (Class Insecta)
      Sucking lice (Order Anoplura)
         Elephant seal louse   *Lepidophthirus macrorhini*
         Leopard, crabeater, Weddell, and Ross seal lice
                             Antarctophthirus spp.
      Flies (Order Diptera)
         Midges (Family Chironomidae)
            Wingless midge   *Belgica antarctica*
            Murphy's midge   *Eretmoptera murphyi*
         Fleas (Order Siphonaptera)
            Silver-gray petrel flea   *Glaciopsyllus antarcticus*
   Crustaceans (Class Crustacea)
      Subclass Malacostraca
         Isopods (Order Isopoda)
            Giant isopod   *Glyptonotus antarcticus*
            Antarcturus isopod   *Antarcturus* sp.

Serolis	*Serolis polita*
	Serolis trilobitoides
Amphipods (Order Amphipoda)	
Giant amphipod	*Bovallia gigantea*
Pontogeniella	*Pontogeniella brevicornis*
Whale louse	*Cyamus* sp.
Krill (Order Euphausiaceae)	
Antarctic krill	*Euphausia superba*
Big-eyed krill	*Thysanoessa* spp.
Ice krill	*Euphausia crystallorophias*
Pygmy krill	*Euphausia frigida*
Northern krill	*Euphausia vallentini*
Spiny krill	*Euphausia triacantha*
Crabs and shrimps (Order Decapoda)	
Lebbeus	*Lebbeus antarcticus*
Copepods (Subclass Copepoda)	
Icefish parasite	*Eubrachiella antarctica*
Sea spiders (Class Pycnogonida)	
Giant sea spider	*Decolopodium antarcticum*
Eight-legged sea spider	*Amothea clausi*
Ten-legged sea spider	*Decolopoda* sp.
Twelve-legged sea spiders	*Dodecolopoda mawsoni*
	Sexanymphon mirabilis
Mollusks (Phylum Mollusca)	
Snails and limpets (Class Gastropoda)	
Antarctic limpet	*Nacella concinna*
Clams (Class Bivalvia)	
Mud clams	*Mysella charcoti*
Nemerteans (Phylum Rhynchocoela)	
Giant nemertean	*Parborlasia corrugatus*
Annelids (Phylum Annelida)	
Leeches (Class Hirudinea)	
Icefish parasite	*Pantobdella rugosa*
Vertebrates (Phylum Vertebrata)	
Tunicates (Subphylum Urochordata)	
Salps (Class Thaliacea)	*Salpa thompsoni*
Bony fishes (Class Osteichthyes)	
Suborder Notothenioidei	

Icefishes (Family Channichthyidae)
    Charcot's icefish   *Parachaenichthys charcoti*
    Gunnar's icefish   *Champsocephalus gunnari*
    White crocodile fish   *Chaenocephalus aceratus*
Antifreeze fishes (Family Nototheniidae)
    Marbled notothenia   *Notothenia rossii marmorata*
    Antarctic toothfish   *Dissostichus mawsoni*
    Pelagic notothenia   *Pleuragramma antarcticum*
Amphibians (Class Amphibia)
    Amphibian (extinct)   Family Labyrinthodontidae
Reptiles (Class Reptilia)
    Aquatic reptile (extinct)   *Lystrosaurus* sp.
Birds (Class Aves)
    Albatrosses (Family Diomedeidae)
        Royal albatross   *Diomedea epomophora*
        Wandering albatross   *Diomedea exulans*
        Black-browed albatross   *Diomedea melanophris*
        Sooty albatross   *Phoebetria fusca*
        Light-mantled sooty albatross   *Phoebetria palpebrata*
    Caciques (Family Icteridae)
        Yellow-rumped cacique   *Cacicus cela*
    Condors (Family Cathartidae)
        Andean condor   *Vultur gryphus*
    Gallinules (Family Rallidae)
        Purple gallinule   *Porphyrula martinica*
        South American terror bird (extinct)   *Titanis* sp.
    Gulls (Family Laridae)
        Southern black-backed gull   *Larus dominicanus*
        Dolphin gull   *Leucophaeus scoresbii*
Penguins (Family Spheniscidae)
    Emperor penguin   *Aptenodytes forsteri*
    Macaroni penguin   *Eudyptes chrysolophus*
    Adélie penguin   *Pygoscelis adeliae*
    Chinstrap penguin   *Pygoscelis antarctica*
    Gentoo penguin   *Pygoscelis papua*
    Magellanic penguin   *Spheniscus magellanicus*
    Galapagos penguin   *Spheniscus mendiculus*
    Jackass penguin   *Spheniscus demersus*

Petrels and prions (Family Procellariidae)
    Antarctic petrel   *Thalassoica antarctica*
    Blue petrel   *Halobaena caerulea*
    Giant petrel   *Macronectes giganteus*
    Pintado petrel   *Daption capense*
    Silver-gray petrel   *Fulmarus glacialoides*
    Snow petrel   *Pagodroma nivea*
    White-chinned petrel   *Procellaria aequinoctialis*
    Antarctic prion   *Pachyptila vittata*
    Narrow-billed prion   *Pachyptila belcheri*
    Southern fulmar   *Fulmarus glacialoides*
Diving petrels (Family Pelacanoididae)
    South Georgia diving petrel   *Pelecanoides georgicus*
Egrets (Family Ardeidae)
    Cattle egret   *Bubulcus ibis*
Pipits (Family Motacillidae)
    South Georgia pipit   *Anthus antarcticus*
Shags and cormorants (Family Phalacrocoracidae)
    Blue-eyed shag   *Phalacrocorax atriceps*
Sheathbills (Family Chionididae)
    American sheathbill   *Chionis alba*
    Lesser sheathbill   *Chionis minor*
Skuas (Family Stercorariidae)
    Brown skua   *Catharacta lonnbergi*
    South polar skua   *Catharacta maccormicki*
Storm petrels (Family Oceanitidae)
    Wilson's storm petrel   *Oceanites oceanicus*
Terns (Family Sternidae)
    Antarctic tern   *Sterna vittata*
    Arctic tern   *Sterna paradisaea*
Waterfowl and swans (Family Anatidae)
    Black-necked swan   *Cygnus melanocoryphus*
Mammals (Class Mammalia)
    Carnivores (Order Carnivora)
        Cats (Family Felidae)
            Domestic cat   *Felis catus*

Marsupials (Order Marsupialia)
  Saber-toothed marsupials (Family Thylacosmilidae)
    Saber-toothed marsupial (extinct)   *Thylacosmilus* sp.
  Possums (Family Didelphidae)
    Seymour Island opposum (extinct)   *Polydolops* sp.
Rodents (Order Rodentia)
  Old World rats and mice (Family Muridae)
    Black rat   *Rattus rattus*
    Norway rat   *Rattus norvegicus*
    House mouse   *Mus musculus*
Rabbits (Order Lagomorpha)
  Rabbits (Family Leporidae)
    Rabbit   *Oryctolagus cuniculus*
Seals and sea lions (Order Pinnipedia)
  Eared seals (Family Otariidae)
    Antarctic fur seal   *Arctocephalus gazella*
    South American fur seal   *Arctocephalus australis*
    South American sea lion   *Otaria byronia*
  True seals (Family Phocidae)
    Crabeater seal   *Lobodon carcinophagus*
    Leopard seal   *Hydrurga leptonyx*
    Elephant seal   *Mirounga leonina*
    Ross seal   *Ommatophoca rossii*
    Weddell seal   *Leptonychotes weddelli*
Whales (Order Cetacea)
  Ancestor of baleen whales (Family Archaeocetae)
     Llanocetus denticrenatus
  Baleen whales (Suborder Mysticetae)
    Rorquals and humpback whale (Family Balaenopteridae)
    Blue whale   *Balaenoptera musculus*
    Fin whale   *Balaenoptera physalus*
    Minke whale   *Balaenoptera acutorostrata*
    Sei whale   *Balaenoptera borealis*
    Humpback whale   *Megaptera novaeangliae*
    Right whales (Family Balaenidae)
    Southern right whale   *Eubalaena australis*

Toothed whales (Suborder Odontoceti)
  Sperm whales (Family Physeteridae)
    Sperm whale   *Physeter macrocephalus*
  Dolphins (Family Delphinidae)
    Long-finned pilot whale
                          Globicephala melaena edwardii
    Orca   *Orcinus orca*
  Bottlenosed and beaked whales (Family Ziphiidae)
    Layard's beaked whale   *Mesoplodon layardii*
    Southern bottlenose whale   *Hyperoodon planifrons*
Even-toed ungulates (Order Artiodactyla)
  Camels (Family Camelidae)
    Guanaco   *Lama guanicoe*
  Sheep (Family Bovidae)
    Sheep   *Ovis aries*
    Mouflon   *Ovis ammon*
  Deer (Family Cervidae)
    Reindeer   *Rangifer tarandus*

Notes

Prologue: Admiralty Bay

1. There is another Banana Belt in Wilkes Land, part of the Indian Ocean sector of Antarctica.
2. Since King George Island, which straddles latitude 62° S, is about 800 kilometers north of the Antarctic Circle (which at 66° 30′ S is the farthest southerly reach of direct sunlight at the June 22 solstice) it has no "midnight sun."
3. Nunatak, an Eskimo word commonly used in Antarctica, means "a hill or mountain completely surrounded by glacial ice" (*Webster's Third New International Dictionary*, 1986).
4. The abbreviation F.I.D.S. in the inscriptions stands for Falkland Islands Dependencies Survey.
5. F.I.D.S. records show that the anonymous memorial is to Ron Napier, who drowned in Admiralty Bay in 1956, when a dinghy, carrying him to shore from the British supply ship *John Biscoe,* overturned. Platt died of a heart attack while conducting a survey near the base. Sharman died while walking with a companion behind the base, when a sled dog knocked them both over a rock cliff. Sharman was killed instantly; his companion was badly bruised but survived. Bell died while surveying a glacier with a dog team. The dogs were tiring in the deep snow and Bell went on ahead, without skis, to encourage them and was swallowed by a crevasse.

He survived the fall of nearly a hundred feet, but when his companion dropped a rope to him, he unwisely tied it to his belt instead of around his waist. Almost at the top of the crevasse, his belt snapped and Bell was never heard from again.

6. Cherry-Garrard 1922, v.
7. Stitching Greenpeace Council 1988, 55.
8. By contrast, the British Antarctic Survey, for example, does not permit women to stay at its stations (although they can work on the ships).

1. Seabirds and Wind

1. In 1986, a 4,144-square-kilometer tabular iceberg was tracked by satellites in the Weddell Sea. It calved so many smaller ice chunks (known, in the vernacular of Antarctic sailors, as "bergy bits" and "growlers") that ships couldn't approach closer than eight kilometers of it. Eventually it drifted northeast into the Scotia Sea, broke apart, and melted.
2. Cook 1900, 122–23.
3. In this book "Antarctic krill" refers only to *Euphausia superba* (see Chapter 5).
4. Among others, these include 44 million diving prions, 8 million diving petrels, 1 million macaroni penguins, 4 million white-chinned petrels, and 1 million plunging and surface-feeding birds.
5. South Georgia, in the Atlantic just south of the Convergence, is biologically Antarctic. It is the home of the only songbird in the Antarctic, the South Georgia pipit, which feeds on insects and seeds, and the South Georgia pintail, which feeds on freshwater vegetation. Kerguelen Island, just north of the Convergence and therefore subantarctic, also has a pintail.
6. The smaller petrels are able to hover, and sometimes appear to walk on water, as did Saint Peter — therefore the name "petrel."
7. Whalers call this oily patch the whale's "footprint."
8. At Britain's Signy Station, in the South Orkney Islands, 524 kilometers northeast of Elephant Island, researchers have reported flocks of up to fifty cattle egrets stranded at once. The cattle egret has perhaps the widest range of any terrestrial bird on Earth. In

this century, it has expanded its breeding range from its native Africa to South America, North America, Europe, Asia and Australia, taking advantage of the burgeoning human population and the concomitant cattle.

2. Memories of Gondwana

1. Eights was the first trained naturalist to visit the Antarctic. During his life he wrote five scientific papers on Antarctic natural history, dealing with subjects as diverse as geology and biology. He was also the first to collect the giant sea spider and the giant amphipod (see Chapter 6).
2. Eights 1838, 206–7.
3. Scott 1913, 440–41.
4. Seymour Island is forty-eight kilometers off the northeastern end of the Antarctic Peninsula in the Weddell Sea. It is only nineteen kilometers long by four to six kilometers wide. What makes it different from other Antarctic islands is that it has almost no permanent ice, a condition that has attracted fossil collectors since the turn of the century. It has also attracted destruction. The Argentinian Marambio Station, with an airstrip that is able to accommodate C-130's, is built on the island. There are two main strata of fossils on the island: Upper Cretaceous sandstones (deposited 80 to 65 million years ago), which contain petrified logs a half meter in diameter, and Upper Eocene sandstones (45 to 40 million years old), which contain the fossils of an extinct two-meter-tall penguin, 40 million years old. The Cretaceous sandstones are particularly interesting to paleontologists because they overlap the mysterious Cretaceous–Tertiary boundary, 66 million years ago, when the dinosaurs became extinct.
5. Quoted in Hooker 1843, 271.
6. A year after his return to England, Ross was married, but his father-in-law set the condition that he never return to sea on long voyages. Ross broke his promise in 1848 when he led a futile attempt to rescue John Franklin's Arctic expedition.
7. Hooker 1843, 280, 328.
8. Weddell 1827, 133.

3. *Life in a Footprint*

1. Lichens are composed of two organisms: a photosynthetic green alga and a fungus (by convention, lichens are named after their fungal component). Under favorable conditions each component can usually live independently of the other, but when joined, they are able to colonize the most inhospitable areas, even bare rock. The mutual benefits of this union are clear. The alga captures the sun's energy and uses it to synthesize carbohydrates; the fungus provides a fibrous matrix that soaks up water. As a team, therefore, they acquire the two most precious resources in Antarctica. Since each lichen has a double genetic endowment, reproduction (when it is not by means of simple fragmentation) is by necessity an elegant synchrony. The fruiting lichen grows asexual, airborne propagules, known as soredia, that have starters of both the alga and the fungus.

2. For the purposes of this discussion, the Antarctic includes Greater Antarctica, Lesser Antarctica, the South Shetlands, the South Orkneys, the South Sandwich Islands, and Bouvetøya Island. All of the remaining terrestrial flora is bacteria (number of species unknown), algae (about 400 species), fungi (about 20 species), lichens (about 150 species, the majority of which are crustose), liverworts (25 species), and mosses (about 75 species). The taxonomies of these groups are in turmoil, for new species are being discovered and old ones merged. In some places the yeasts, many of which are epiphytic on mosses, comprise about 10 percent of the microbial biomass.

3. Bundle grass, the southernmost flowering plant in the world, is found as far south as the Terra Firma Islands of Marguerite Bay (68° 42′ S), on the west side of the Antarctic Peninsula.

4. Only four phyla of terrestrial invertebrates, an impoverished and often scrofulous fauna, manage to eke out a living in Antarctica. There are approximately 65 species of protozoans in the maritime Antarctic and South Georgia, 40 species of nematode worms (of which 34 are endemic), two annelids, and of the arthropods, 17 known species of tardigrades, 8 species of springtails, and 67 species of insects (including 2 species of midges, 32 species of biting lice, and 3 of sucking lice). As with the Antarctic flora, the

discovery of new species continues and their taxonomies are being continually revised. The exact number of taxa is therefore unknown.

5. There is a winged midge, *Parochlus steineni*, in the Antarctic Peninsula and the nearby islands. But its ability to fly is facultative; adults may be full-winged, partially winged, or wingless. The specific environmental conditions that influence the expression of wings are unknown.

6. But since Antarctica is a land of predictable climate, however harsh, an opposite mating strategy is also adaptive, as illustrated by a winged relative of *Belgica*, Murphy's midge, *Eretmoptera murphyi*. A native of South Georgia that was accidentally introduced to Signy Island, Murphy's midges are all females. The species is probably parthenogenetic; that is, females manufacture fertile eggs without mating, creating virtual clones of themselves. This strategy affords reproductive certainty, but none of the genetic experimentation of sex.

7. Hooker 1843, pp. 247, 326.

8. Quoted in Hooker 1843, pp. 262–63, 287.

9. Quoted in Hooker 1843, p. 258.

10. The governor of the Falkland Islands reported to Hooker that two American sailors, shipwrecked in the West Falklands, had subsisted for fourteen months solely on the roots of tussock grass. The sailors even made their home beneath a clump of this grass, using a round boulder as a door.

11. Quoted in Hooker 1843, pp. 280, 302.

12. Ross and Hooker attempted several other interesting plant introductions. They transported hundreds of southern beeches from Patagonia to the Falkland Islands in an effort to turn those treeless islands into an arbor (and, unknown to them, restore the genus to the islands where it had prospered millions of years before). They also took Kerguelen cabbage to Tasmania, where it was planted in the garden of Governor John Franklin (who was later to perish on a celebrated expedition to the Arctic). The plants thrived there, but in subsequent years did not survive the long voyage across the torrid zones to England.

13. Quoted in Hooker 1843, p. 327.

14. Hooker 1843, p. xi.

4. Penguins and Hormones

1. Webster 1834, vol. 1, pp. 144, 148–49.
2. Webster 1834, vol. 1, p. 163.
3. Webster 1834, vol. 1, pp. 159–60.
4. Webster 1834, vol. 1, p. 149.
5. Some penguin species acquired their vernacular and Latin names in interesting ways. The Adélie was named by the nineteenth-century French explorer Jules-Sébastien-César Dumont d'Urville after his wife. History does not record whether she was as squat and frumpy as her namesake. The macaroni penguin is named for its crest of feathers, which to the eyes of the early Antarctic explorers resembled the foppish coiffure of members of London's Macaroni Club. Gentoo is a name first used by English explorers in the Falkland Islands and is the archaic word for "Hindu," referring to the twin triangular white patches, reminiscent of a turban, on the penguin's head. The Latin name for the gentoo, *Pygoscelis papua*, is one of the great misnomers of science. The bird's describer, Johann Reinhold Forster (after whom the emperor penguin, *Aptenodytes forsteri*, is named), a naturalist who circumnavigated the world with Captain James Cook, falsely believed that the species occurred in Papua New Guinea. In fact, the nearest gentoos to New Guinea are on Macquarie Island, 6900 kilometers due south.

   The remaining eleven species of penguins live in cold waters of the temperate and even tropical latitudes. Only one, the Galápagos penguin, is equatorial, but its survival depends on the cold Humboldt Current, which sweeps north from the Southern Ocean. The Galápagos Islands are also bathed by warm equatorial currents, making them the only place on Earth where penguins may be seen swimming over tropical coral reefs. The Galápagos penguin is at the extreme limit of its physiological tolerance, and in El Niño years, when warm equatorial surface water pushes aside the Humboldt Current and accumulates in the eastern Pacific, the species suffers reproductive failure and decline.
6. This research has been conducted since 1976 by Nick Volkman, Wayne and Susan Trivelpiece, and their colleagues at the Point

Reyes Bird Observatory (Trivelpiece, Trivelpiece, and Volkman, 1987).

7. Cherry-Garrard 1922, p. 268.

8. But Cherry-Garrard lived to write a book on Scott's last expedition, including the journey to Cape Crozier. It was aptly titled *The Worst Journey in the World,* and may be the finest account of Antarctic exploration ever written.

9. Indeed, the South African jackass penguin is named for its call.

10. A second skua species, the south polar skua, is also found in the South Shetland Islands. It is so closely related to the brown skua that in certain areas the two species have hybridized. The south polar skua ranges farther south than any vertebrate species except humans; individuals have been seen at the South Pole. The species also raids penguin colonies, but in the vicinity of the Antarctic Peninsula, perhaps because of competition from the more abundant brown skua, it feeds mostly at sea.

11. However, the south polar skuas, which eat krill and fish, fare poorly during El Niño years.

12. The tame skuas that frequent research stations, waiting for handouts of spoiled food and garbage, often have a lower reproductive success than the wild skuas, which have a high-quality diet from fishing or patrolling the penguin rookeries.

13. There are two species of sheathbills in the Antarctic. The lesser sheathbill lives in the subantarctic islands of the Indian Ocean and does not reach the continent or its fringing islands.

5. The Galaxies and the Plankton

1. On the night of February 23–24, 1987, exactly one year before I wrote these words, a supernova, named 1987A, appeared in the Large Magellanic Cloud. It was the first supernova observed in the neighborhood of Earth since the ones described by Tycho Brahe in 1572 and Johannes Kepler in 1604. At its peak three days later, supernova 1987A was approximately 100 million times brighter than the Earth's sun and was visible to the naked eye. The explosion also created a pulse of neutrinos, subatomic particles so

small they can pass through objects the size of planets without colliding with other matter. Approximately eighteen hours before visible light from the supernova reached Earth, neutrinos were detected simultaneously in Ohio and in Japan by scientists working in subterranean laboratories. The mass of neutrinos has long been a subject of fierce debate among particle physicists. One group argued that their mass was relatively large, indeed, that neutrinos accounted for much of the unseen dark matter of the universe. Another group argued that neutrinos were ghostly particles without any mass at all. The debate was unexpectedly resolved by supernova 1987A. Because the neutrinos from the explosion reached Earth before the photons, physicists deduced that their mass must be close to, if not actually, zero.

Supernovas are the source of most of the heavy elements in the universe. Most of the silicon that makes up the rocks of Keller Peninsula and the shells of diatoms in Admiralty Bay originated in supernovas similar to 1987A.

2. There are about one hundred species of diatoms in Antarctic waters. Twenty are endemic; two are bipolar, occurring in both the Arctic and the Antarctic.

3. The auxospore is not a true spore in the sense of dormantly enduring harsh environmental conditions.

4. The potential lifespan of Antarctic krill is a subject of intense, and controversial, scientific inquiry. If humans ever harvest krill on a large scale, this information will be essential for the proper management of the resource. It is nearly impossible to determine the age of an adult krill because its size does not necessarily correlate with its age. Biologists have long assumed that crustaceans periodically shed their confining old shells in order to grow larger. However, recently it has been discovered that during the winter, when diatoms are relatively scarce in the Southern Ocean, Antarctic krill may actually shrink during a molt and regress from sexually mature to sexually immature as a means of economizing scarce resources. This means, for example, that a sexually immature krill 1.5 centimeters long may be older than a mature one 2.0 cm long. It may be possible to get around this problem by measuring the levels of the pigment lipofuscin in the tissues of krill. Lipofuscin is a metabolic by-product that accumulates at a steady

rate over a krill's lifetime; its concentration may be a function of a krill's age.

5. The swarm would not have been detected from the surface of the sea, except that more than forty Russian trawlers were fishing it.

6. Most krill-eating species of squids in the Southern Ocean have lifespans of only a year or two. But the krill that they prey on can live up to seven years, creating the unusual circumstance of a predator being shorter-lived than its prey.

7. But such a harvest would, of course, affect the animals that feed on krill, particularly in lean years. For example, in 1977–78 the usual huge swarms of krill failed to appear in the waters around South Georgia and that summer many of the seabirds that depend on krill failed to reproduce. If during those years krill had been harvested by humans on a grand scale, then the effects on the seabirds might have been catastrophic.

8. Hooker 1847, p. 503.

6. *The Bottom of the Bottom of the World*

1. Eights 1833, cited in Hedgpeth 1971b, p. 13.

2. Eights 1835, cited in Hedgpeth 1971b, p. 19.

3. However, not all tunicates are sessile. Salps, which occur in astronomical numbers in the Antarctic plankton, are free-floating colonial tunicates. Like their sessile relatives, they filter plankton and nutrients from the water.

4. But here the similarities between the Southern Ocean and the Arctic Ocean end. The Southern Ocean, a part of the Atlantic, Pacific, and Indian oceans, yet isolated by currents and wind at the bottom of the world, surrounds a continent. The Arctic Ocean is surrounded by continents, and because of the massive effluvia of the Siberian and North American rivers that debouch into it, is in fact an estuary.

5. Amundsen began his trek from an indentation in the northeastern edge of the Ross Ice Shelf known as the Bay of Whales. Ever a meticulous planner, he pondered the risk of setting up camp on floating ice. He wrote:

The ice barrier — so much described in all works on Antarctic exploration — is in reality nothing but a gigantic glacier pressing down from the heights of the Antarctic mountains to the sea. This glacier is hundreds of kilometers in width and from thirty to sixty meters high. Like all glaciers, at its lower end this one was constantly breaking off into icebergs. The idea, therefore, of making a permanent camp on the barrier itself, though often considered, had always been dismissed as too dangerous.

I had, however, carefully read and long pondered the works of the earlier explorers in the Antarctic. In comparing their records, I had been greatly struck with the discovery that the Bay of Whales, notwithstanding that it was merely a bay whose shores were the icy walls of the glacier, had not substantially changed its shore line since its first discovery by Sir James Ross in 1842. "Surely," I said to myself, "if this part of the glacier has not moved in sixty-eight years, there can only be one explanation of the phenomenon — the glacier at this point must be firmly wedged on the solid rock of some great and immovable island." The more I thought of this explanation, the more I became convinced of its truth. I had, therefore, no fear of the stability of our camp site when I resolved to make our permanent land quarters on the top of the barrier in the Bay of Whales. Needless to say, my faith was entirely justified by subsequent events. We had the most delicate instruments and we made continuous observations for months, none of which disclosed the slightest movement of the barrier of ice at this point. [Amundsen 1927, pp. 68–69]

Amundsen was just lucky. In October 1987 a tabular iceberg, ignominiously named B-9, measuring 155 by 35 kilometers (about the size of Long Island, New York) and bearing the entire Bay of Whales, cleaved off the Ross Ice Shelf. B-9 contained 287 cubic miles of fresh water, enough to provide every person on Earth with two glasses of water per day for 1,977 years. Satellites and aircraft tracked it for two years. By August 1989 it had drifted to the northwest, crashed into the coast of Victoria Land, rebounded, and broken into three pieces. By July 1990 the three pieces were still visible off the coast of George V Land, 806 kilometers west of Cape Adare.

6. Not surprisingly, the benthic fauna of western McMurdo Sound, which borders the Ross Ice Shelf, is depauperate, sustaining low population densities of only about 350 species. By contrast, the oxygenated eastern sound has faunal densities as high as any place on Earth: up to 140,000 individuals (of innumerable species) per square meter.

7. Recently there has been speculation that these buried lakes may lubricate the passage of the ice above them, and that if greenhouse warming of Earth's atmosphere softens the ice shield, the lakes may hasten its slide toward the ocean, raising sea levels all over the world.

7. The Worm, the Fish and the Seal

1. Eights 1846, cited in Hedgpeth 1971b, pp. 5–6 (see References, Chapter 6).

2. Because Admiralty Bay is diluted by melting ice, the freezing temperature of the surface waters, depending on proximity to the glaciers, is a few tenths of a degree higher than that of the open sea.

3. The few species of notothenids, such as *Pleuragramma,* that are pelagic and feed on the abundant krill and plankton have adopted a different life strategy from that of their bottom-dwelling relatives. They grow much faster, mature at an earlier age, and have greater fecundity, producing up to 100,000 eggs, which are dispersed into the plankton.

    Obviously, bottom-dwelling notothenids cannot sustain a large-scale fishery. Yet in response to the imposition of 200-mile exclusive fishing zones in territorial waters in the north, the world's fisheries have turned to the stocks of fish in the largely unclaimed Southern Ocean. Small-scale trials using purse seines and bottom trawls were initiated as early as the 1930s by Argentinian and Norwegian companies, but the easy profits to be made by whaling in the same waters made the enterprises less attractive. During the early 1960s, as whaling became less economical due to the decimation of stocks, fishing in the Southern Ocean was resumed, particularly by the Russians in the vicinity of South Georgia Island, where the marbled notothenia, which spends its early years in the fjords feeding on benthic fish and invertebrates, was particularly abundant. The annual catch slowly increased until the summer of 1970–71, when 366 million kilograms of marbled notothenia were taken. The population soon collapsed, and the following season only 10 million kilograms were caught; during the summer of 1972–73 none were taken. By then, fisheries biologists, hobbled by a lack of even rudimentary knowledge of the populations, growth patterns, and fecundity of notothenid fish, made their first estimate of what would constitute a sustainable harvest. It was only 45 million kilograms per year for *all* species of bottom-dwelling fish, which were anonymously designated "unspecified demersal percomorphs" by the Antarctic Treaty members. It was therefore no wonder that ten years after the disastrous 1970–71 season, the Scotia Sea population of marbled notothenia had still not recovered.

4. The leopard seals in Admiralty Bay are intensely curious about humans and commonly swim around the Zodiacs, arching their heads high to peer into the boat. They have, to date, never attacked a human in Admiralty Bay. However, scuba divers in other parts of the Southern Ocean have reported attacks by leopard seals. In one remarkable instance in February 1974, biologists T. DeLaca and G. Zumwalt, scuba diving in Arthur Harbor near the American Palmer Station on Anvers Island, were harassed for forty-five minutes by a leopard seal. The divers retreated to a rocky depression at a depth of five meters, where the seal laid siege to them for thirty minutes. Like a skua, the seal attacked from the direction of the sun. The divers defended themselves with pieces of angle iron that they found on the seafloor, but the seal repeatedly struck the irons. The divers were rescued when the Zodiac driver managed to distract the seal.

5. Elephant seals have been hunted for their oil and fur seals for their pelts in the Antarctic and subantarctic since the 1700s (see Chapters 8 and 9). Fur sealing began on South Georgia in the 1780s. By 1802 it was estimated that 112,000 skins had been taken from the island. The final fur-sealing expedition was in 1913, but the species was by then so rare that only 170 skins were taken, and as recently as 1930 there were only a few hundred fur seals on the

island. Only in recent decades has the population of fur seals begun to recover (and dramatically, with annual rates of population increase close to 17 percent). The elephant seal, which was less profitable to hunt than the fur seal, did not suffer as drastic a decline, and by 1910 there were sufficient numbers of sealers on South Georgia that the newly established British administration on the island began to issue licenses for controlled hunting of the species. The hunt ended in 1965 (at the same time as the cessation of whaling on South Georgia — see Chapter 11). Although it was a large-scale process, with groups of sealers sailing around the island and taking as many as 6,000 elephant seals per year, there was no significant decline in their numbers. The reason for this excellent stewardship was an applied understanding of the social structure of the seals. Only the males were hunted, and since most males were reproductively irrelevant, the population growth rate did not diminish (indeed, given the diminished competition for food resources because of the removal of the males, it may have actually increased). This success was primarily due to the inspiration of Richard Laws, a versatile mammalogist who also conducted pioneering studies on elephants in Uganda and who eventually became the director of the British Antarctic Survey. Laws instructed the sealers to take a tooth from 5 percent of the animals that they killed. By counting the number of rings laid down in the dentin, each of which represented one of two annual fasting haulouts, Laws was able to calculate the ages of the sampled seals and to estimate the age structure of the population, and the annual harvest was adjusted accordingly.

8. Visions of Ice and Sky

1. Cited in Christie 1951, p. 30.
2. Fletcher's relationship with Drake was as tempestuous as the seas they crossed. At one point Drake had Fletcher placed in irons, and inscribed on his arm, "frances [sic] fletcher ye falsest knave yt. liveth." (Mill 1905, p. 33.)
3. But Drake's crew were not the first Europeans to observe pen-

guins. That distinction fell to the crew of Portuguese navigator Bartholomeu Diaz on November 22, 1497, near the Cape of Good Hope. Diaz wrote, "There are birds as big as ducks, but they cannot fly, because they have no feathers on their wings. These birds, of whom we killed as many as we chose, are called Fotylicayos, and they bray like asses" (Simpson, p. 3). Diaz was describing jackass penguins, close relatives of the Magellanic penguin, both of which indeed bray like asses.

4. Cited in Christie 1951, p. 37.
5. Cited in Reader's Digest 1985, p. 69.
6. Cited in Christie 1951, p. 41.
7. The Kerguelen Islands are situated just north of the Antarctic Convergence. Geographically, therefore, they are only marginally Antarctic, and their flora and fauna are typically subantarctic.
8. Cited in Mill 1905, p. 52.
9. Reader's Digest 1985, p. 73.
10. Cited in Migot 1956, p. 11.
11. Beaglehole 1961, p. 168.
12. Furneaux, an accomplished explorer in his own right, became the first person to circumnavigate the world in both directions.
13. Cook 1777, vol. 1, p. 56.
14. Beaglehole 1961, pp. 304–5.
15. Cook 1777, vol. 1, pp. 253–54.
16. The Little Ice Age ended at about the same time that modern weather records started being maintained. Most of our information on the period is indirect, from the analysis of tree rings. From these data it is estimated that mean temperatures in the North Atlantic during the period were between 1° and 3° C lower than the present.
17. Cited in Christie 1951, p. 71.
18. Miers 1820, p. 369
19. Cited in Christie 1951, p. 72.
20. The elephant seal populations proved more resilient to human predation than the fur seal populations, even though from 1850 to 1860 American sealers brought home 1,976,751 gallons of elephant seal oil, and the hunting of elephant seals on South Georgia persisted for over a century longer.
21. Christie 1951, p. 72.
22. Cited in Christie 1951, p. 74.

23. Cited in Christie 1951, p. 76.
24. Young 1821, p. 345.
25. The *Hersilia* soon encountered the Shag Rocks, four bird-guan-oed mountain tops 185 kilometers west-northwest of South Georgia. These rocks, only about thirty meters tall, are easy to miss in the open sea. Steep-sided, with jagged tips, they protrude like daggers from open ocean; looming and vanishing in the sea fog, they were probably the source of the Aurora Island myth. But lacking beaches and therefore having no seals, the rocky islands proved worthless to Sheffield, and the *Hersilia* turned back to the Falklands.
26. Young 1821, p. 346.
27. Young 1821, p. 346.
28. Young 1821, p. 347.
29. Mill 1905, p. 98.

9. *The Indifferent Eye of God*

1. Byers Peninsula is now designated a specially protected area under the terms of the Antarctic Treaty.
2. Quoted in Balch 1909, pp. 481–82. The original spelling and abbreviations of this manifest, as duly recorded by Balch, have been retained. However, certain items that were ambiguous or partially missing have been omitted.
3. This dubious distinction has been claimed by others. In 1876–77, a second party was forced to overwinter on King George Island at Potter Cove, 24 kilometers southwest of Admiralty Bay. A gang of sealers from the *Florence*, of New London, Connecticut, under the command of the mate, a Mr. King of Yonkers, New York, was dropped at Rugged Island, about 130 kilometers southwest of Potter Cove, but because of the unexpected drifting of the pack ice, could not be picked up. After sailing to Potter Cove, they made camp for the winter under their upturned dinghy. All died except King, who was rescued the following summer.

   In 1898–99 the Belgian exploring ship *Belgica*, with a polyglot crew under the command of Adrien de Gerlache, was trapped in the pack ice of the Bellingshausen Sea, southwest of King George

Island, and had to overwinter. Locked in the ice, the ship drifted hundreds of kilometers. During the winter darkness, one of the nineteen expedition members died, two went insane, and almost everyone suffered from fear and depression.

That expedition inaugurated the epic age of polar exploration. The *Belgica*'s first mate was the Norwegian Roald Amundsen, who was the first person to sail the Northwest Passage (1903–1906); in 1911 he became the first person to reach the South Pole. The physician on the *Belgica* was an American, Frederick Cook, who later led a controversial expedition toward the North Pole. Among the scientists on the *Belgica* were a Pole, Henryk Arcktowski, who pioneered Antarctic meteorology; a Rumanian, Emile Racovitza, who described several new species; and a Belgian, Emile Danco, who conducted some of the first studies of magnetism in the Antarctic. Cook's beautifully written account of this voyage, *Through the First Antarctic Night* (1900), is a classic of exploration writing. As implied by the title, the crew of the *Belgica* believed they were the first party to overwinter in the Antarctic; in fact, they were the third.

4. Fildes 1821. His account is the first record of an alien animal in the Antarctic. Today a menagerie of feral animals, deliberately or accidentally introduced, has created a serious environmental problem on islands in the Antarctic and subantarctic. They include the Norway rat (South Georgia), the black rat (Macquarie and Kerguelen), the house mouse (South Georgia, Macquarie, Kerguelen, Marion, Crozet), the rabbit (Macquarie, Kerguelen, Crozet), the sheep (Kerguelen Island), the reindeer (South Georgia, Kerguelen, and Crozet), the mouflon (Kerguelen), and the trout (Kerguelen Island).

The domestic cat has been introduced to Kerguelen, Marion, Macquarie, and Crozet islands. On Kerguelen, cats annually kill about 1,200,000 birds, mostly prions. During the summer of 1949–50, 5 cats were introduced to Marion Island. By 1965 their population was approximately 2,000, ranging from the coastal plain to the snow line, and that year it was estimated that the cats killed about 455,000 petrels. During March 1977, in an effort to control the feline population, 96 cats infected with feline distemper virus were released on the island by the South African government. By 1988 the cat population had fallen to less than 100.

5. Clark 1887, p. 437.
6. Smith 1844, pp. 159–61.
7. To this day, things haven't improved. The most recent conflict in the Southern Ocean was the 1982 war between Great Britain and Argentina over the Falkland Islands and South Georgia. The fundamental reason for this war was the Treaty of Tordesillas, which in 1494 gave to Spain all South American territories (including the Falklands) west of 47° W, and to which, as successors to the Spanish Empire, the Argentines to this day consider themselves the legitimate heirs. Another reason for the conflict was that these islands may sit atop untapped strata of natural gas. The British claim to the Antarctic is independent of their claim to the Falklands, although as a matter of convenience they administer their Antarctic Territory from the Falklands.
8. Quoted in Balch 1909, p. 481. This list has been edited for brevity and chronological order.
9. Unpublished logbook of the schooner *Huntress,* cited in Bertrand 1971, p. 97.
10. Many years later, Palmer told the tale to one Mr. Bush, the American consul in Hong Kong, who transcribed it from memory. Cited in Balch 1909, pp. 479–80.
11. The dates of Bellingshausen's journey have been converted from the "old system" calendar to their equivalents in the modern calendar.
12. Bellingshausen 1831, vol. 1, pp. 127–28.
13. Bellingshausen 1831, vol. 2, p. 419.
14. Quoted in Balch 1909, p. 480.
15. Bellingshausen 1831, vol. 2, pp. 425–26.
16. Weddell 1827, pp. 141–42.
17. Weddell 1827, p. 137.
18. Weddell 1827, p. 143.

10. *The Tern and the Whale*

1. Quoted in Christie 1951, p. 139 (see References, Chapter 8).
2. Ross 1847, vol. 1, p. 191.
3. Cook 1900, p. 148 (see References, Chapter 1).

4. From the Latin for "mustached," a reference to the baleen.
5. The maximum lengths and weights of the mysticete whales are as follow: blue whale (30 meters, 135,000 kilograms), fin whale (25 meters, 82,000 kilograms), sei whale (18 meters, 27,000 kilograms), minke whale (11 meters, 17,000 kilgrams), humpback whale (16 meters, 64,000 kilograms) and right whale (18 meters, 82,000 kilograms). The lengths and weights of the odontocete whales are: sperm whale (males 18 meters, 64,000 kilograms; females 11 meters 15,000 kilograms), orca (males 9 meters, 7,000 kilograms; females 8 meters, 5,500 kilgrams), southern bottlenose whale (7 meters, 5,000 kilograms), long-finned pilot whale (5 meters, 900 kilograms) and Layard's beaked whale (5 meters, 900 kilograms). The males of the sperm whale and the orca are larger than the females, indicating that they are harem animals.
6. Quoted in Winn and Winn 1985, p. 91.
7. By contrast, cow's milk is 4 percent fat and 16 percent protein.
8. Orcas are commonly known as killer whales, an unfair moniker that invites fear and B-movie scenarios. No wild killer whale has ever attacked a human. Indeed, humans are by far the most ruthless killers in the Antarctic, and if the standards we apply to "killer whales" were applied to us, we would deserve the name of "killer hominids."
9. Cooperation by males in mating has also been observed in mysticete whales, including the right whale.
10. Villiers 1925, pp. 195–96.
11. The blow is angled because the toothed whales have only one, rather crooked, nostril located on the left side of the head. The right branch of their respiratory tract is closed. (The mysticete whales, by contrast, have two symmetrical nostrils.)

11. *The Passing of the Leviathans*

1. *New York Times,* Feb. 15, 1981, p. 26. As the ship began to sink, the crew hastily provisioned three open whaleboats with food, fresh water, knives, and pistols. A month later the three boats reached uninhabited Henderson Island, where the sailors took

on more fresh water and dried fish. Three of the crewmen stayed on the island and were eventually rescued, but the others, including Nickerson, continued eastward. Storms soon separated the boats. By January 18, one of the sailors on Nickerson's boat, Richard Peterson, had died, and his body was cast into the sea. But when a second sailor, Isaac Cole, went insane and died on February 8, rations were so short that the flesh of his limbs was stripped off and eaten by the three survivors. Nine days later Nickerson and his two companions were rescued by a passing vessel. A second boat, with the captain and a surviving crew of one, was also eventually rescued. The third was lost at sea with all hands.

2. Malone 1854, pp. 69–71.
3. Donald 1893, p. 438.
4. Murdoch 1894, pp. 269–70.
5. Murdoch 1894, pp. 289–90.
6. Murdoch 1894, p. 295.
7. Larsen 1894, pp. 333, 337.
8. Cited in Reader's Digest 1985, p. 152 (see References, Chapter 8).
9. Eighty years later Seymour Island was the site of the discovery of *Polydolyps,* the first mammalian fossil in Antarctica.
10. Tønnesen and Johnsen 1982, pp. 284–85.
11. Larsen's petition to the British government also had tremendous political significance. It was one of the first acknowledgments of British sovereignty over South Georgia and set a precedent for all territorial claims in the Antarctic.
12. Bagshawe 1939, pp. 8–9.
13. Villiers 1925, pp. 98–99.
14. Villiers 1925, p. 49.
15. Villiers 1925, pp. 123–24.
16. Villiers 1925, p. 150.
17. Bagshawe 1939, pp. 7–8.
18. Villiers 1925, pp. 132–33.
19. Bagshawe 1939, p. 24.
20. Villiers 1925, p. 136.
21. Bennett 1931, p. 42.
22. Bagshawe 1939, pp. 7–8.

23. Bagshawe 1939, p. 24.
24. Villiers 1925, p. 100.
25. Today the Church of the Snows is farther south, in McMurdo Sound.
26. Tønnesen and Johnsen 1982, p. 286.
27. Quoted in Tønnesen and Johnsen 1982, p. 290.
28. Tønnesen and Johnsen 1982.
29. Until recently, the scientific study of the great whales has been like the study of the smallest subatomic particles: biologists couldn't get information about them without permanently altering the subject. In the case of whales, this meant killing them, usually as a by-product of the whaling industry. Almost nothing, therefore, is known about their behavior on the high seas. Yet from this destructive analysis, cetologists have been able to glean the essential details of the natural history of whales, such as their age structures and rates of reproduction.

    For centuries humans believed that because whales grow to such huge sizes, they must live a long time, like the giant redwoods. But we now know that whales live no longer than people. How do you age a whale? In mysticete whales, age can be determined by counting the number of layers in a waxy, ligamentous extension of the eardrum, which is periodically shed. In toothed whales, age is estimated by counting the number of dentine rings in the teeth. However, since few whales of known age have been dissected, there has been controversy as to the number of dental rings and lamellae that are deposited every year, so the data can be grossly misleading. In fin whales, for example, early studies concluded that two lamellae were laid down every year. However, subsequent studies on tagged fin whales showed that only one was laid down per annum. Similarly, in sperm whales, the first studies indicated that two dentine rings were laid down per year; now it is widely accepted that the number is one. By the brutally wasteful method of dissecting pregnant female whales, their breeding season and the rate of fetal growth have been estimated. A female whale's reproductive history is revealed by the number of scars etched in her ovaries. A follicular scar, known as the corpus albicans, is made each time she ovulates. A larger scar, known as the corpus luteum, is created every time she conceives, the result of the production of progesterone, a hormone that enriches the lining of the uterus. By counting the scars of each type, biologists are able to determine her mating success.

12. *The Tempest*

1. The twelve were Argentina, Australia, Belgium, Chile, France, Great Britain, Japan, New Zealand, Norway, South Africa, the United States, and the USSR.
2. In addition to the twelve charter nations, there are thirteen additional consultative signatories: Brazil, China, Finland, Germany, India, Italy, Peru, Poland, South Korea, Spain, Sweden, Switzerland, and Uruguay. The remaining members are: Austria, Bulgaria, Canada, Colombia, Cuba, Czechoslovakia, Denmark, Ecuador, Greece, Hungary, the Netherlands, North Korea, Papua New Guinea, and Romania.
3. Greenpeace soon began to bring world attention to the heaps of garbage at the nearby American base at McMurdo Sound and at New Zealand's Scott Base. In petty retaliation, the personnel of Scott Base, who oversee the maintenance of Scott's hut (located in the New Zealand Antarctic Protectorate), refused to allow the Greenpeace personnel to enter the hut until February 1988, when they were permitted to accompany a group of tourists from the *World Discoverer*.
4. Exploratory drilling for oil and natural gas offshore in the Falkland Islands is expected to begin in 1992. By 1991, British Petroleum, Shell, Occidental Petroleum, and Conoco had all started negotiations for licenses with the British government. In the meantime, Britain is guarding the Falklands, which has a resident population of only 2,121, with approximately 2,000 troops.
5. Wilkes and Mann 1978, p. 34.

References

The research described in this book is the result of hard work by scientists and explorers who have devoted many years — and in some cases their lives — to understanding the Antarctic. The following sources are the primary and secondary references that were used for each chapter.

Prologue: Admiralty Bay

Cherry-Garrard, Apsley. 1922. *The Worst Journey in the World*. One-volume edition of 1937. London: Chatto and Windus.

Defense Mapping Agency Hydrographic/Topographic Center. 1985. *Sailing Directions (Planning Guide and Enroute) for Antarctica*. Washington, D.C.: Defense Mapping Agency, Office of Distribution Services.

Fuchs, Vivian. 1982. *Of Ice and Men: The Story of the British Antarctic Survey, 1943–73*. Oswestry, England: Anthony Nelson.

Hattersley-Smith, G. 1951. "King George Island." *Alpine Journal* 58(282): 67–76.

Headland, R. K., and P. L. Keage. 1985. "Activities on the King

George Island Group, South Shetland Islands, Antarctica." *Polar Record* 22(140): 475–84.

Noble, H. M. 1965. "Glaciological observations at Admiralty Bay, King George Island, in 1957–58." *British Antarctic Survey Bulletin* 5: 1–11.

Stitching Greenpeace Council. 1989. *Expedition Report 1987–88, Greenpeace Antarctic Expedition.* East Sussex, England.

Vtyurin, B. I., and M. Yu. Moskalevskiy. 1985. "Cryogenic Landforms, King George Island, South Shetland Islands." *Polar Geography and Geology* 9: 62–69.

1. Seabirds and Wind

Cook, Frederick A. 1900. *Through the First Antarctic Night, 1898–1899.* London. Reprint, 1980. Montreal: McGill–Queen's University Press.

Croxall, J. P. 1984. "Seabirds." In R. M. Laws, ed. *Antarctic Ecology,* vol. 2, pp. 533–616. London: Academic Press.

Jameson, William. 1959. *The Wandering Albatross.* New York: William Morrow and Co.

Jouventin, Pierre, and Henri Weimerskirch. 1990. "Satellite Tracking of Wandering Albatrosses." *Nature* 343: 746–48.

Siegfried, W. R. 1985. "Birds and Mammals — Oceanic Birds of the Antarctic." In W. N. Bonner and D. W. H. Walton, eds. *Key Environments, Antarctica,* pp. 242–65. Oxford: Pergamon Press.

Watson, George E. 1975. *Birds of the Antarctic and Sub-Antarctic.* Washington, D.C.: American Geophysical Union.

2. Memories of Gondwana

Adie, Raymond J. 1964. "Geological History." In Raymond Priestly, Raymond J. Adie, and G. de Q. Robin, eds. *Antarctic Research,* pp. 117–62. London: Butterworth.

Anonymous. 1985. "Descubrimiento de un antigo mamífero fósil."
    Antartida 14: 14–21.

Bakker, E. M. van Zinderen. 1970. "Quaternary Climates and Ant-
    arctic Biogeography." In M. W. Holdgate, ed. *Antarctic Ecol-
    ogy*, pp. 31–40. New York: Academic Press.

Eights, James. 1838. "A Description of the New South Shetland
    Islands." In Edmund Fanning. *Voyages to the South Seas, Indian
    and Pacific Oceans, China Sea, North-West Coast, Feejee
    Islands, South Shetlands, &c.*, pp. 193–216. Reprint, 1970.
    Saddle River: Gregg Press.

Jefferson, Timothy H., and Thomas N. Taylor. 1983. "Permian and
    Triassic Woods from the Transantarctic Mountains: Paleoenvi-
    ronmental Indicators." *Antarctic Journal of the United States
    1983 Review:* 55–57.

John, B. S., and D. E. Sugden. 1971. "Raised Marine Features and
    Phases of Glaciation in the South Shetland Islands." *British
    Antarctic Survey Bulletin* 24: 45–111.

Scott, Robert Falcon. 1913. *Scott's Last Expedition.* London: Smith,
    Elder and Co.

Stubblefield, S. P., and T. N. Taylor. 1985. "Fossil Fungi in Antarctic
    Wood." *Antarctic Journal of the United States 1985 Review:* 7–8.

Weddell, James. 1827. *A Voyage Toward the South Pole Performed
    in the Years 1822–24, Containing an Examination of the Ant-
    arctic Sea.* 2nd ed. Reprint. Devon, England: David and Charles
    Reprints.

Zinsmeister, William J. 1984. "Geology and Paleontology of Sey-
    mour Island, Antarctic Peninsula." *Antarctic Journal of the
    United States* 19(2): 1–5.

———. 1986. "Fossil Windfall at Antarctica's Edge." *Natural His-
    tory* 95(5): 60–67.

3. Life in a Footprint

Block, W. 1984. "Terrestrial Microbiology, Invertebrates and Eco-
    systems." In R. M. Laws, ed. *Antarctic Ecology*, vol. 1, pp. 163–
    236. London: Academic Press.

Cranston, P. S. 1985. "*Eretmoptera murphyi* Schaeffer (Diptera:

Chironomidae), an Apparently Parthenogenetic Antarctic Midge." *British Antarctic Survey Bulletin* 66: 35–45.

Friedmann, E. Imre. 1982. "Endolithic Microorganisms in the Antarctic Cold Desert." *Science* 215: 1045–53.

Gressitt, J. Linsley. 1965. "Terrestrial Animals." In Trevor Hatherton, ed. *Antarctica*, pp. 351–71. New York: Frederick A. Praeger.

Hooker, T. N. 1980a. "Factors Affecting the Growth of Antarctic Crustose Lichens." *British Antarctic Survey Bulletin* 50: 1–19.

———. 1980b. "Growth and Production of *Cladonia rangiferina* and *Sphaerophorus globosus* on Signy Island, South Orkney Islands." *British Antarctic Survey Bulletin* 50: 27–34.

———. 1980c. "Growth and Production of *Usnea antarctica* and *U. fasciata* on Signy Island, South Orkney Islands." *British Antarctic Survey Bulletin* 50: 35–49.

Hooker, William J. 1843. "Notes on the Botany of H. M. Discovery Ships, *Erebus* and *Terror* in the Antarctic Voyage; with Some Account of the Tussac Grass of the Falkland Islands." *London Journal of Botany* 2: 247–329.

Light, J. J., and R. B. Heywood. 1973. "Deep-water Mosses in Antarctic Lakes." *Nature* 242: 535–36.

———. 1975. "Is the Vegetation of Continental Antarctica Predominantly Aquatic?" *Nature* 256: 199–200.

Longton, R. E. 1985. "Terrestrial Habitats—Vegetation." In W. Bonner and D. W. H. Walton, eds. *Key Environments, Antarctica*, pp. 73–105. Oxford: Pergamon Press.

Redon, Jorge. 1985. *Liquenes Antarcticos*. Santiago: Instituto Antarctico Chileno.

Smith, R. I. Lewis. 1981. "The Earliest Report of a Flowering Plant in the Antarctic?" *Polar Record* 20: 571–72.

———. 1984. "Terrestrial Plant Biology." In R. M. Laws, ed. *Antarctic Ecology*, vol. 1, pp. 61–162. London: Academic Press.

———. 1985. "*Nothofagus* and Other Trees Stranded on Islands in the Atlantic Sector of the Southern Ocean." *British Antarctic Survey Bulletin* 66: 47–55.

Sømme, Laurits. 1985. "Terrestrial Habitats — Invertebrates." In W. Bonner and D. W. H. Walton, eds. *Key Environments, Antarctica*, pp. 73–105. Oxford: Pergamon Press.

4. Penguins and Hormones

Ainley, David G. 1975. "Displays of Adélie Penguins: A Reinterpretation." In Bernard Stonehouse, ed. *The Biology of Penguins,* pp. 503–34. London: Macmillan.

Ainley, David G., Robert E. LeResche, and William J. L. Sladen. 1983. *Breeding Biology of the Adélie Penguin.* Berkeley: University of California Press.

Cherry-Garrard, Apsley. 1922. *The Worst Journey in the World.* One-volume edition of 1937. London: Chatto and Windus.

Conroy, J. W. H. 1975. "Recent Increases in Penguin Populations in Antarctica and the Subantarctic." In Bernard Stonehouse, ed. *The Biology of Penguins,* pp. 321–36. London: Macmillan.

Croxall, J. P. 1984. "Sea Birds." In R. M. Laws, ed. *Antarctic Ecology,* vol. 2, pp. 533–616. London: Academic Press.

Müller-Schwarze, C., and D. Müller-Schwarze. 1975. "A Survey of Twenty-four Rookeries of Pygoscelid Penguins in the Antarctic Peninsula Region." In Bernard Stonehouse, ed. *The Biology of Penguins,* pp. 309–20. London: Macmillan.

Müller-Schwarze, D. 1984. *The Behavior of Penguins, Adapted to Ice and Tropics.* Albany: State University of New York Press.

Poncet, S., and J. Poncet. 1987. "Censuses of Penguin Populations of the Antarctic Peninsula, 1983–87." *British Antarctic Survey Bulletin* 77: 109–29.

Simpson, George Gaylord. 1976. *Penguins, Past and Present, Here and There.* New Haven: Yale University Press.

Spurr, E. B. 1975. "Communication in the Adélie Penguin." In Bernard Stonehouse, ed. *The Biology of Penguins,* pp. 449–501. London: Macmillan.

Stonehouse, B. 1985. "Birds and Mammals—Penguins." In W. Bonner and D. W. H. Walton, eds. *Key Environments, Antarctica,* pp. 266–92. Oxford: Pergamon Press.

Trivelpiece, Susan G., and Wayne Z. Trivelpiece. 1989. "Antarctica's Well-bred Penguins." *Natural History* Dec. 1990: 29–36.

Trivelpiece, Wayne Z. 1987. "Breeding Penguins and the Marine Food Web, Austral Summer." *Point Reyes Bird Observatory Newsletter* 79: 1–5.

Trivelpiece, Wayne Z., Susan G. Trivelpiece, and Nicolas J. Volkman.

1987. "Ecological Segregation of Adélie, Gentoo, and Chinstrap Penguins at King George Island, Antarctica." *Ecology* 68(2): 351–61.

Watson, George E. 1975. *Birds of the Antarctic and Sub-Antarctic.* Washington, D.C.: American Geophysical Union.

Webster, W. H. B. 1834. *Narrative of a Voyage to the Southern Atlantic Ocean in the Years 1828, 29, 30, Performed in H. M. Sloop Chanticleer Under the Command of the Late Captain Henry Foster, F.R.S. & 'c. by Order of the Lords Commissioners of the Admiralty.* 2 vols. London. Facsimile, 1970. London: Dawsons of Pall Mall.

5. The Galaxies and the Plankton

Adelung, D., K. Kössmann, and D. Rössler. 1985. "The Distribution of Fluoride in Some Antarctic Seals." *Polar Biology* 5: 31–34.

Culik, Boris. 1987. "Fluoride Turnover in Adélie Penguins *(Pygoscelis adeliae)* and Other Bird Species." *Polar Biology* 7: 179–87.

George, Robert Y. 1980. "Pressure and Temperature Adaptations of Antarctic Krill and Common Peracarid Crustaceans." *Antarctic Journal of the United States* 15(5): 145–46.

———. 1981. "Euphausiid Larval Distribution in the Scotia Sea, 1979–1980." *Antarctic Journal of the United States* 16(5): 141–42.

Golosov, V. V. 1984. "Some Biocenological Aspects of the Exploitation of the Resources of Antarctic Krill." *Vestnik Leningradskogo Universiteta* (1980) 12: 95–103. Translation in *Polar Geography and Geology* 8: 63–72.

Hamner, William M. 1982. "Procedures for *In Situ* Observations of Krill Schools in the Southern Ocean." *Antarctic Journal of the United States* 17(5): 165.

Hooker, Joseph D. 1847. *The Botany of the Antarctic Voyage of H. M. Discovery Ships* Erebus *and* Terror *in the Years 1839–1843.* Vol. 1. Reprint, 1963. Weinheim, Germany: J. Cramer.

Macaulay, Michael. 1981. "Distribution and Abundance of Krill in the Scotia Sea as Observed Acoustically, 1981." *Antarctic Journal of the United States* 16(5): 166–67.

————. 1983. "Antarctic Krill *(Euphausia superba)* Swarms from Elephant Island." *Antarctic Journal of the United States* 18(5): 197–98.

Marr, James. 1962. "The Natural History and Geography of the Antarctic Krill *(Euphausia superba* Dana)." *Discovery Reports* 32.

Miller, D. G. M., and I. Hampton. 1989. *Biology and Ecology of the Antarctic Krill* (Euphausia superba *Dana): A Review.* Cambridge, England: Scott Polar Research Institute.

Morris, D. J., and J. Priddle. 1984. "Observations on the Feeding and Moulting of the Antarctic Krill, *Euphausia superba* Dana, in Winter." *British Antarctic Survey Bulletin* 65: 57–63.

Nicol, Stephen. 1990. "The Age-old Problem of Krill Longevity." *BioScience* 40(11): 833–36.

Ochlenschläger, J., and M. Manthey. 1982. "Fluoride Content of Antarctic Marine Animals Caught off Elephant Island." *Polar Biology* 1: 125–27.

Quetin, Langdon B., and Robin M. Ross. 1983. "Larval Growth and Hatching under Pressure of Eggs of *Euphausia superba*." *Antarctic Journal of the United States* 18(5): 200–202.

Schneppenheim, R., and C. M. MacDonald. 1984. Genetic Variation and Population Structure of Krill *(Euphausia superba)* in the Atlantic Sector of Antarctic Waters off of the Antarctic Peninsula." *Polar Biology* 3: 19–28.

Schulenberger, Eric. 1983. "Superswarms of Antarctic Krill *(Euphausia superba* Dana)." *Antarctic Journal of the United States* 18(5): 194–97.

Waldrop, M. Mitchell. 1987. "Sighting of a Supernova." *Science* 235: 1143.

Wormuth, John H. 1983. "Zooplankton Associated with Superswarms of Antarctic Krill." *Antarctic Journal of the United States* 18(5): 199–200.

6. The Bottom of the Bottom of the World

Amundsen, Roald. 1927. *My Life as an Explorer.* Garden City, N.Y.: Doubleday, Page and Co.

Bruchhaiusen, P. M., et al. 1979. "Fish, Crustaceans, and the Sea Floor under the Ross Ice Shelf." *Science* 203: 449–51.

Dell, R. K. 1965. "Marine Biology." In Trevor Hatherton, ed. *Antarctica,* pp. 129–52. New York: Frederick A. Praeger.

Eights, James. 1833. "Description of a New Animal Belonging to the Crustacea." *Transactions of the Albany Institute* 2: 331–34.

———. 1835. "Description of a New Animal Belonging to the Arachnides of Latreille." *Boston Journal of Natural History* 1(2): 203–6.

———. 1846. "On the Icebergs of the Ant-Arctic Sea." *American Quarterly Journal of Agriculture and Science* 4(1): 20–24.

Hedgpeth, Joel W. 1971a. "Perspectives in Benthic Ecology in Antarctica." In Louis O. Quam and Horace D. Porter, eds. *Research in the Antarctic,* pp. 93–136. Washington, D.C.: American Association for the Advancement of Science.

———. 1971b. "James Eights of the Antarctic (1798–1882)." In Louis O. Quam and Horace D. Porter, eds. *Research in the Antarctic,* pp. 3–45. Washington, D.C.: American Association for the Advancement of Science.

Hessler, Robert R. 1970. "High-Latitude Emergence of Deep-sea Isopods." *Antarctic Journal of the United States* July-Aug. 1970: 133–34.

Hooker, Joseph D. 1847. *The Botany of the Antarctic Voyage of H. M. Discovery Ships* Erebus *and* Terror *in the Years 1839–1843.* Vol. 1. Reprint, 1963. Weinheim, Germany: J. Cramer.

Lipps, J. H., T. E. Ronan, and T. E. DeLaca. 1979. "Life below the Ross Ice Shelf, Antarctica." *Science* 203: 447–49.

Luxmore, R. A. 1982. "The Reproductive Biology of Some Serolid Isopods from the Antarctic." *Polar Biology* 1: 3–11.

McClintock, James B. 1985. "Avoidance and Escape Responses of the Sub-Antarctic Limpet *Nacella edgari* (Powell) (Mollusca: Gastropoda) to the Sea Star *Anasterias perrieri* (Smith) (Echinodermata: Asteroidea)." *Polar Biology* 4: 95–98.

Menzies, Robert J., and Robert Y. George. 1969. "Polar Faunal Trends Exhibited by Antarctic Isopod Crustacea." *Antarctic Journal of the United States* Sept.-Oct. 1969: 190–91.

Picken, Gordon B. 1985. "Marine Habitats—Benthos." In W. Bonner and D. W. H. Walton, eds. *Key Environments, Antarctica,* pp. 154–72. Oxford: Pergamon Press.

Richardson, Michael G., and Terence M. Whitaker. 1979. "An Antarctic Fast-Ice Food Chain: Observations on the Interaction of the Amphipod *Pontogeneia anatarctica* Chevreux with Ice-Associated Micro-algae." *British Antarctic Survey Bulletin* 47: 107–15.

Walker, A. J. M. 1972. "Introduction to the Ecology of the Antarctic Limpet *Patinigera polaris* (Hombron and Jacquinot) at Signy Island, South Orkney Islands." *British Antarctic Survey Bulletin* 28: 49–69.

White, M. G. 1984. "Marine Benthos." In R. M. Laws, ed. *Antarctic Ecology,* vol. 2, pp. 421–61. London: Academic Press.

7. The Worm, the Fish, and the Seal

Bagshawe, Thomas Wyatt. 1939. *Two Men in the Antarctic: An Expedition to Graham Land 1920–1922.* New York: Macmillan.

DeLaca, T. E., J. H. Lipps, and G. S. Zumwalt. 1975. "Encounters with Leopard Seals *(Hydruga leptonyx)* along the Antarctic Peninsula." *Antarctic Journal of the United States* 10(3): 85–91.

Eastman, Joseph T. 1980. "Evolutionary Divergence in McMurdo Sound Fishes." *Antarctic Journal of the United States 1980 Review:* 151–53.

Eastman, Joseph T., and Arthur L. DeVries. 1986. "Antarctic Fishes." *Scientific American* 255(5): 106–14.

Everson, I. 1968. "Larval Stages of Certain Antarctic Fishes." *British Antarctic Survey Bulletin* 16: 65–70.

———. 1978. "Antarctic Fisheries." *Polar Record* 19(120): 233–51.

———. 1984. "Fish Biology." In R. M. Laws, ed. *Antarctic Ecology,* vol. 2, pp. 491–532. London: Academic Press.

Kock, Karl-Hermann. 1985. "Marine Habitats — Antarctic Fish." In W. Bonner and D. W. H. Walton, eds. *Key Environments, Antarctica,* pp. 173–92. Oxford: Pergamon Press.

Kooyman, Gerald L. 1981. *Weddell Seal, Consummate Diver.* Cambridge: Cambridge University Press.

Ruud, Johan T. 1965. "The Ice Fish." *Scientific American* 213(5): 108–14.

Stackpole, Edouard A. 1953. *The Sea-Hunters, New England Whalemen During Two Centuries.* Philadelphia: J. B. Lippincott.

8. Visions of Ice and Sky

Beaglehole, J. C., ed. 1961. *The Journals of Captain James Cook on his Voyages of Discovery: The Voyage of the* Resolution *and* Adventure *1772–1775.* Hakluyt Society. Cambridge: Cambridge University Press.

Chernavskaya, M. M. 1985. "Reconstruction of Temperature During the Little Ice Age in Northern Eurasia, Based on Dendrochronological Data." *Polar Geography and Geology* 9: 321–28.

Cherry-Garrard, Apsley. 1922. *The Worst Journey in the World: Antarctic 1910–1913.* One-vol. edition of 1937. London: Chato and Windus.

Christie, E. W. Hunter. 1951. *The Antarctic Problem.* London: George Allen and Unwin.

Cook, James. 1777. *A Voyage Towards the South Pole and Round the World,* 2 vols. Dublin: W. Whitstone.

Drake, Francis. 1628. *The World Encompassed.* Reprint, 1966. Readex Microprint Corp.

Miers, J. 1820. "Account of the Discovery of New South Shetland, with Observations on Its Importance in a Geographical, Commercial, and Political Point of View." *Edinburgh Philosophical Journal* 3: 367–80.

Migot, André. 1956. *Thin Edge of the World.* Boston: Little, Brown.

Mill, Hugh R. 1905. *The Siege of the South Pole.* New York: Frederick A. Stokes.

Owen, Russell. 1941. *The Antarctic Ocean.* New York: Whittlesey House.

Reader's Digest. 1985. *Antarctica: Great Stories from the Frozen Continent.* Sydney, Australia: Reader's Digest.

Rubin, Morton J. 1982. "James Cook's Scientific Programme in the Southern Ocean, 1772–75." *Polar Record* 21(130): 33–49.

Simpson, George Gaylord. 1976. *Penguins, Past and Present, Here and There*. New Haven: Yale University Press.

Smith, G. Barnett. 1900. *The Romance of the South Pole*. London: Thomas Nelson and Sons.

Stehberg, Rubén L. 1983. "En torne a la autenticidad de las puntas de proyectil aborigenes descubiertas en las Islas Shetland Sur." *Boletín Antartico Chileno* 3(1): 21–22.

Young, Adam [incorrectly attributed to "H. M. S. Slaney"]. 1821. "Notice of the Voyage of Edward Barnsfield, Master of His Majesty's Ship Andromache to New South Shetland." *Edinburgh Philosophical Journal* 4: 345–48.

9. The Indifferent Eye of God

Balch, Edwin Swift. 1901. "Antarctica: A History of Antarctic Discovery." *Journal of the Franklin Institute* 151(4): 241–62; 151(5): 321–428; 152(1): 26–45.

———. 1904. "Antarctic Addenda." *Journal of the Franklin Institute* 157(2): 81–88.

———. 1909. "Stonington Antarctic Explorers." *Bulletin of the American Geographical Society* 41(8): 473–92.

Bellingshausen, Thaddeus von. 1931. *The Voyage of Captain Bellingshausen to the Antarctic Seas 1819–1821*. 2 vols. English translation, 1945. Frank Debenham, ed. London: Hakluyt Society.

Bertrand, Kenneth J. 1971. *Americans in Antarctica, 1775–1948*. New York: American Geographical Society.

Clark, A. Howard. 1887. "The Antarctic Fur-seal and Sea-elephant Industry." In George Brown Goode, ed. *The Fisheries and Fishery Industries of the United States*, pp. 400–67. Washington, D.C.: Government Printing Office.

Eights, James. 1838. "A Description of the New South Shetland Islands." In Edmund Fanning. *Voyages to the South Seas, Indian and Pacific Oceans, China Sea, North-West Coast, Feejee Islands, South Shetlands, & c.*, pp. 193–216. Reprint, 1970. Saddle River: Gregg Press.

Fildes, Robert. 1821. "Journal of a Voyage Kept on Board Brig Cora of Liverpool, Bound for New South Shetland, 1820–21." Typescript copy made for W. S. Bruce in 1916 of original logs in the Public Record Office (shelf mark: Admiralty, ships' logs 143, supplementary series 11). Cambridge, England: Scott Polar Research Institute Library.

Leader-Williams, N. 1985. "The Sub-Antarctic Islands — Introduced Mammals." In W. N. Bonner and D. W. H. Walton, eds. *Key Environments, Antarctica*, pp. 318–28. Oxford: Pergamon Press.

Miers, J. 1820. "Account of the Discovery of New South Shetland, with Observations on Its Importance in a Geographical, Commercial, and Political Point of View." *Edinburgh Philosophical Journal* 3: 367–80.

Murdoch, W. G. Burn. 1894. *From Edinburgh to the Antarctic: An Artist's Notes and Sketches During the Dundee Antarctic Expedition of 1892–93*. London: Longmans, Green and Co.

Roberts, Brian. 1958. "Chronological List of Antarctic Expeditions." *Polar Record* 9(59): 97–134; 9(60): 191–239.

Smith, R. I. Lewis, and H. W. Simpson. 1987. "Early Nineteenth-Century Sealers' Refuges on Livingston Island, South Shetland Islands." *British Antarctic Survey Bulletin* 74: 49–72.

Smith, Thomas W. 1844. *Narrative of the Life, Travels and Sufferings of Thomas W. Smith*. New Bedford, Mass.: William C. Hill.

Stackpole, Edouard A. 1953. *The Sea-Hunters: The New England Whalemen During Two Centuries, 1635–1835*. Philadelphia: J. B. Lippincott.

Webster, W. H. B. 1834. *Narrative of a Voyage to the Southern Atlantic Ocean in the Years 1928, 29, 30*. 2 Vols. Facsimile, 1970. London: Dawsons of Pall Mall.

Weddell, James. 1827. *A Voyage Toward the South Pole Performed in the Years 1822–24, Containing an Examination of the Antarctic Sea*. 2nd ed. Reprint, 1970. Devon, England: David and Charles Reprints.

Young, Adam [incorrectly attributed to "H. M. S. Slaney"]. 1821. "Notice of the Voyage of Edward Barnsfield, Master of His Majesty's Ship Andromache, to New South Shetland." *Edinburgh Philosophical Journal* 4: 345–48.

10. *The Tern and the Whale*

Darling, J. D., K. M. Gibson, and G. K. Silber. 1983. "Observations on the Abundance and Behavior of Humpbacked Whales *(Megaptera novaeangliae)* off West Maui, Hawaii, 1977–79." In R. Payne, ed. *Communication and Behavior of Whales,* pp. 210–22. Boulder, Colo.: Westview Press.

Davies, J. L. 1963. "The Antitropical Factor in Cetacean Speciation." *Evolution* 17: 107–16.

Gambell, Ray. 1968. "Seasonal Cycle and Reproduction in Sei Whales of the Southern Hemisphere." *Discovery Reports* 35: 35–131.

———. 1985. "Birds and Mammals — Antarctic Whales." In W. N. Bonner and D. W. H. Walton, eds. *Key Environments, Antarctica,* pp. 223–41. Oxford: Pergamon Press.

Glockner, D. A., and S. C. Venus. 1983. "Identification, Growth Rate, and Behavior of Humpback Whale *(Megaptera novaeangliae)* Cows and Calves in the Waters off Maui, Hawaii, 1977–79." In R. Payne, ed. *Communication and Behavior of Whales,* pp. 223–58. Boulder, Colo.: Westview Press.

Heimlich-Boran, S. L. 1986. "Cohesive Relationships among Puget Sound Killer Whales." In B. C. Kirkevold and J. S. Lockard, eds. *Behavioral Biology of Killer Whales,* pp. 251–84. New York: Alan R. Liss.

Jacobsen, J. K. 1986. "The Behavior of *Orcinus orca* in the Johnstone Strait, British Columbia." In B. C. Kirkevold and J. S. Lockard, eds. *Behavioral Biology of Killer Whales,* pp. 135–85. New York: Alan R. Liss.

Laws, R. M. 1959. "The Foetal Growth Rates of Whales, with Special Reference to the Fin Whale *Balaenoptera physalis* Linn." *Discovery Reports* 29: 283–308.

———. 1961. "Reproduction, Growth and Age of Southern Fin Whales." *Discovery Reports* 31: 327–486.

———. 1981. "Experiences in the Study of Large Mammals." In Charles W. Fowler and Tim D. Smith, eds. *Dynamics of Large Mammal Populations,* pp. 19–45. New York: John Wiley and Sons.

Lockard, J. S. 1986. "Research Status of *Orcinus orca:* What Is Not Known about Its Behavioral Biology." In B. C. Kirkevold and

J. S. Lockard, eds. *Behavioral Biology of Killer Whales,* pp. 407–42. New York: Alan R. Liss.

Mackintosh, N. A., and J. F. G. Wheeler. 1929. "Southern Blue and Fin Whales." *Discovery Reports* 1: 257–540.

Nickerson, Thomas. Journal quotes, *New York Times,* June 1981.

Payne, K. 1983. "Progressive Changes in the Songs of Humpback Whales *(Megaptera novaeangliae):* A Detailed Analysis of Two Seasons in Hawaii." In R. Payne, ed. *Communication and Behavior of Whales,* pp. 9–57. Boulder, Colo.: Westview Press.

Purves, P. E., and G. E. Pilleri. 1983. *Echolocation in Whales and Dolphins.* London: Academic Press.

Ross, James Clark. 1847. *A Voyage of Discovery and Research in the Southern and Antarctic Regions During the Years 1839–43.* Reprint, 1969. Devon, England: David and Charles Reprints.

Stone, G. S., and W. M. Hamner. 1988. "Humpback Whales *Megaptera novaeangliae* and Southern Right Whales *Eubalaena australis* in Gerlache Strait, Antarctica." *Polar Record* 24(148): 15–20.

Townsend, C. H. 1935. "The Distribution of Certain Whales as Shown in Logbook Records of American Whaleships." *Zoologica* 19(1): 1–50.

Villiers, A. J. 1925. *Whaling in the Frozen South: Being the Story of the 1923–24 Norwegian Whaling Expedition to the Antarctic.* Indianapolis: Bobbs-Merrill.

Watkins, W. A., and D. Wartzok. 1985. "Sensory Biophysics of Marine Mammals." *Marine Mammal Science* 1(3): 219–60.

Winn, Lois King, and Howard E. Winn. 1985. *Wings in the Sea: The Humpbacked Whale.* Hanover, N.H.: University Press of New England.

Würsig, B. 1989. "Cetaceans." *Science* 244:1550–57.

Zagaeski, M. 1987. "Some Observations on the Prey Stunning Hypothesis." *Marine Mammal Science* 3(3): 275–79.

11. The Passing of the Leviathans

Bagshawe, Thomas Wyatt. 1939. *Two Men in the Antarctic: An Expedition to Graham Land, 1920–1922.* New York: Macmillan.

Bennett, A. G. 1931. *Whaling in the Antarctic*. Edinburgh: William Blackwood and Sons.

Breiwick, J. M., and H. W. Braham, eds. 1984. "The Status of Endangered Whales." *Marine Fisheries Review* 46(4).

Brown, S. G., and C. H. Lockyer. 1984. "Whales." In R. M. Laws, ed. *Antarctic Ecology*, vol. 2, pp. 717–81. London: Academic Press.

Bruce, W. S. 1893. "A Voyage toward the Antarctic Sea, September 1892 to June 1893. Preliminary Report [on *Balaena*]." *Geographical Journal of London* 2: 430–33.

Cherfas, Jeremy. 1986. "What Price Whales?" *New Scientist*, June 5, 1986.

Cook, Frederick A. 1900. *Through the First Antarctic Night, 1898–1899*. London. Reprint, 1980. Montreal: McGill–Queen's University Press.

Cook, James. 1777. *The Voyage Towards the South Pole and Round the World*. 2 Vols. Dublin.

Donald, C. W. 1893. "A Voyage Toward the Antarctic Sea, September 1892 to June 1893. Preliminary Report [on *Active*]." *Geographical Journal of London* 2: 433–38.

Fraser, Conon. 1986. *Beyond the Roaring Forties: New Zealand's Subantarctic Islands*. Wellington, N.Z.: Government Printing Office.

Hayes, J. Gordon. 1928. *Antarctica: A Treatise on the Southern Continent*. London: Richards Press.

Headland, R. K., and P. L. Keage. 1985. "Activities on the King George Island Group, South Shetland Islands, Antarctica." *Polar Record* 22(140): 475–84.

Jackson, Gordon. 1978. *The British Whaling Trade*. Hamden, Conn.: Archon Books, Shoe String Press.

Larsen, C. A. 1894. "The Voyage of the 'Jason' to the Antarctic regions." *Geographical Journal of London* 4: 333–44.

Malone, R. Edmond. 1854. *Three Years' Cruise in the Australasian Colonies*. London: Richard Bentley.

Murdoch, W. G. Burn. 1894. *From Edinburgh to the Antarctic: An Artist's Notes and Sketches During the Dundee Antarctic Expedition of 1892–93*. London: Longmans, Green and Co.

Murphy, Robert Cushman. 1947. *Logbook for Grace, Whaling Brig Daisy, 1912–1913*. New York: Time.

Ross, James Clark. 1847. *A Voyage of Discovery and Research in the Southern and Antarctic Regions During the Years 1839–43.* Reprint, 1969. Devon, England: David and Charles Reprints.

Smith, G. Barnett. 1900. *The Romance of the South Pole: Antarctic Voyages and Explorations.* London: Thomas Nelson and Sons.

Tønnessen, J. N. 1970. "Norwegian Antarctic Whaling, 1905–68: An Historical Approach." *Polar Record* 15(96): 283–90.

Tønnessen, J. N., and A. O. Johnsen. 1982. *The History of Modern Whaling.* Berkeley: University of California Press.

Villiers, A. J. 1925. *Whaling in the Frozen South: Being the Story of the 1923–24 Norwegian Whaling Expedition to the Antarctic.* Indianapolis: Bobbs-Merrill.

12. The Tempest

Barnaby, Frank. 1989. "Antarctica: The First of Five Nuclear Weapon Free Zones." *Ambio* 18(1): 90–91.

Marshall, Eliot. 1980. "Scientists Fail to Solve Vela Mystery." *Science* 207: 504–6.

———. 1980. "Debate Continues on the Bomb That Wasn't." *Science* 209: 572–73.

———. 1980. "Navy Lab Concludes the Vela Saw a Bomb." *Science* 209: 996–97.

Wilkes, Owen, and Robert Mann. 1978. "The Story of Nukey Poo." *Bulletin of the Atomic Scientists* Oct. 1978: 32–36.

Index